EVERYMAN, I will go with thee,

and be thy guide,

In thy most need to go by thy side

DAVID RICARDO

Born in 1772, the son of a Dutch Jew. Made
a fortune on the London Stock Exchange.
Retired to Gloucestershire, 1814. M.P. from
1819 until death in 1823.

DAVID RICARDO

The Principles
of Political Economy
and Taxation

INTRODUCTION BY
MICHAEL P. FOGARTY,
HON. D.SC. POL. SOC. SC. (LOUVAIN)
Professor of Industrial Relations, University of Wales

DENT: LONDON
EVERYMAN'S LIBRARY
DUTTON: NEW YORK

NO. 590

SBN: 460 00590 1

INTRODUCTION

DAVID RICARDO was born in 1772. His father, an orthodox Jew originally from Holland, had settled in England and made his fortune on the Stock Exchange. Before David was well in his teens he began to know his way around the financial world. By the age of twenty-one he had married, and also left the Jewish faith, which meant separating from his father. But he struck out for himself on the Stock Exchange, and before he was thirty had made a large fortune of his own. With this he bought an estate in Gloucestershire, and in 1819 a seat in Parliament. He died in 1823, after what was on the surface a successful but uneventful life; uneventful, that is, considering that this was the age of the French Revolution, Napoleon, and the Revolutions in agriculture and industry. He was:

'... a good husband and father, a man kindly, modest, and unassuming, without artifice or pretension, in discussion more ready to listen than to speak, frank in acknowledging error and in admitting conviction, and at the same time quietly cogent and compelling in the advance and illustration of his own conclusions.' F. W. Kolthammer, original introduction to the *Principles of Political Economy and Taxation*, Everyman ed., p. ix.

Ricardo first became interested in economic theory in 1799, through reading Adam Smith's *Wealth of Nations*. Through the next ten years he studied, at first casually, then more deeply, until in 1810, following some correspondence in the *Morning Chronicle*, he wrote his first pamphlet on *The High Price of Bullion*. This impressed the experts, and brought him a good deal of publicity. He began to correspond with the leading economists and political writers of the day, including Malthus, Bentham, Say, and James Mill. Much of his thought and findings in economics went into this correspondence, a large part of which has been published for the first time only within the present generation. In 1930, when Lord Keynes had set on foot a scheme for a definitive edition of Ricardo's works, a search was started for the important series of letters then still missing. A large box of letters received by Ricardo was quickly discovered by one of his descendants in the house

formerly owned by his eldest son. Little by little other finds
were added. There remained one major gap; Ricardo's letters
to James Mill, which happen to be particularly important for
the origins of the *Principles of Political Economy and Taxation*.
For years the search went on among descendants of Mill and
their executors and friends all over the world, until at last, in
1943, the gap was filled. A box containing the missing letters
was found in the house near Dublin formerly owned by J. E.
Cairnes, the economist, a close friend of James Mill's son John
Stuart Mill, through whom presumably he came into possession
of them.

With the help of these recent discoveries the story of the
Principles can be put together. Ricardo was never a willing
writer. Much of his best work had to be in effect dragged out of
him by his friends. The same was true of him as a speaker.
One of the earlier of his few speeches in Parliament was made
only when he was called on by members on all sides of the
House. The *Principles* were no exception. The chief driving
force in this case seems to have come from James Mill, who was
anxious to see Ricardo state more fully the principles under-
lying the *Essay on the Influence of a Low Price of Corn on the
Profits of Stock* which he published early in 1815. The scheme
took shape through 1815. By 1816 it was well under way, and
in October of that year Ricardo sent Mill a draft covering what
are now the first seven chapters of the *Principles*, that is the
chapters concerned with basic theory. By the middle of
November he sent a draft of the chapters on taxation. He
then spent a couple of months re-reading works by Adam
Smith, Say, and Malthus, and drafting the chapters at the end
of the *Principles* in which some of these authors' positions are
criticized. The printing of the book was begun at the end of
February 1817, and it was published on 19th April. A second
and third edition, both revised by Ricardo himself, appeared in
1819 and 1821. It is on the third and final revision that the
present edition is based.

Even before he wrote the *Principles*, Ricardo was recognized
as the outstanding British economist of his day. The *Principles*
themselves have had an immense and world-wide influence.
Karl Marx was as much Ricardo's disciple as Hegel's. Marx
stood Ricardo, like Hegel, on his head before using him; the
conclusions of Marx's *Capital* are not precisely what Ricardo
would have wished. But the fact remains that *Capital* is the

pure milk of Ricardo's doctrine, developed and interpreted by a brilliant disciple, with a fire and venom and skill in practical illustration which the master himself never equalled. Among non-Marxist economists Ricardo has been read and re-read down to the present day. As a statement of current economic thought the *Principles* have long been obsolete. Since Jevons and Menger laid the foundations of marginal analysis, economics has come to state even those truths grasped by Ricardo in a terminology unknown to him. And he did not grasp by any means the whole truth of economics, nor even all its main lines. But the *Principles* remain one of the great documents of economic thought, with a place in history alongside Smith's *Wealth of Nations* or Malthus's *Principle of Population*. We read them now not for any new lessons they might teach us but as a record of one of the great pioneers, on whose limited and one-sided achievements has been built the more complex and comprehensive knowledge of to-day.

How much then did Ricardo grasp and achieve? It is not always easy to say. Being a graceless as well as an unwilling writer, he often conceals his meaning or scatters it in bits and pieces through his work. He will say in his Preface that the principal problem of political economy is that of sharing the national income between rent, profit, and wages. But towards the end of the book he will make clear that he merely means that this is the problem hardest for economists to sort out. What matters most in the working of the economic system, he now explains, is the volume of production and the absolute income received by each person concerned in it. Or he will talk about the rise or fall of real wages: and it suddenly dawns on one that he is giving this phrase a meaning which to modern readers will seem unnatural and distorted. For us to-day it means a rise or fall in the purchasing power of wages. But for him it meant a rise or fall in the percentage share of wages in the national income. Or again, at first meeting he seems to be a downright upholder of a labour theory of value. But read on, and this theory turns out to be qualified to a point where a modern economist need find nothing to quarrel with in it. Profits figure in some passages as 'a residual'; they are what remains to the farmer, for instance, 'after the landlord and labourer are paid.' But elsewhere it appears, correctly, that profits are the supply price of the services rendered by capitalists and entrepreneurs. But when one has penetrated the smoke-screen of Ricardo's style, one finds that he is astonishingly often right. There are great, glaring, gaps. But in what he actually says it is not so easy to catch him out.

(1) *Ricardo's theory of economic control.*

A modern economist might begin by asking what Ricardo has to say about economic control. Suppose that consumers and savers are agreed as to what the economic system should produce. How effective are the various mechanisms of economic control in ensuring that it gets produced? What will each of them achieve? Let us translate Ricardo's reply into modern economic language. Full, effective, or perfect competition, he says, will cause goods to be sold at a price equal to marginal cost in the case of farm products and to both average and marginal cost in that of manufactures. The distinction arises because he treats farm products as being produced typically under conditions of rising marginal cost— hence the existence of rent—whereas manufactures are produced under constant costs. Marginal cost, Ricardo would say, includes 'normal' wages, profits, and depreciation. Wages or profits, like prices, are 'normal' or competitive when they are equal to the marginal cost of supplying the services in question; to the cost of the customary standard of life of the marginal worker of each grade, and the cost of an acceptable allowance for the risk and 'waiting' of the marginal investor. There are points here where a modern economist would wish to be more explicit, or perhaps even to correct minor errors. But as a general statement of what might be expected to happen under effective or perfect competition what Ricardo says passes well enough.

Only there is so much that he does not say; so much ground that he does not cover. An economist to-day would wish to press a great deal harder on the question of what happens when competition is less than perfect. How often is it imperfect, in what ways, to what extent, with what results? How often, for instance, does one find cases where goods or services are produced under conditions not of constant but of diminishing costs, so that there is a tendency for the size of plants or departments to expand until there is a high degree of oligopoly, or even a case of monopoly? And even when competition is perfect, are there not many things which one would not expect it to achieve? Can it be expected to enforce correct decisions in cases where many of the results or costs to be taken into account are so remote in time or so widely diffused as to be unknown or not immediately important to the people who, under competition, would have to take the decision? What about the trade cycle, or town planning, or the survival and development of local communities as such? Or what about those aspects of the distribution of income and wealth not immediately obvious in, say, an individual wage bargain, and therefore liable to be overlooked when one is made? Or can one be as sure as Ricardo sometimes seems to be that competition will offer rewards sufficient to induce pioneers to innovate, breaking through existing social processes and structures on the way? Competition *compels* conformity to existing standards, but merely offers the *opportunity*, and some inducement, to advance into new fields. Is there not such a thing as a 'competition of dullards,' in which all conform to existing standards but no one goes beyond them?

Ricardo did not ignore such points as these. He was perfectly aware that partial or total monopoly is possible. He did not deny that some economic functions are beyond the scope of individuals and of competition and need to be performed collectively. He showed himself aware, in his chapters on 'Sudden Changes in the Channels of Trade' and 'On Machinery,' of at least some of the snags which may arise when structural change in an economy is carried out by competition alone. But under all these headings his touch is less sure and his analysis much more sporadic than in his discussion of effective or perfect competition.

This is also true of another question which preoccupies modern economists. If competition partly or totally breaks down, what is to take its place? A modern text-book cannot avoid dealing at length with the problems of economic direction by the State, or public corporations, or large private concerns. It must also discuss consultative procedures in the firm, on an industry-wide basis, between Government and industry, and on the international level. And it must show how to choose the appropriate mechanism in each case, and to police its operation. Ricardo has some useful things to say in this field, particularly in his long discussion of taxes and subsidies and their effect on the flow of trade and the volume of savings. But much the greater part of what an economist to-day would say about these matters is not to be found in Ricardo at all.

(2) *Ricardo's theory of the choice of objectives.*

The problem of economics is not of course simply to discover how given decisions about what is desirable are to be carried out. It is also to show how these decisions are themselves arrived at, and can be influenced. What leads people to consume what they do, to invest what they do, to accept a greater or less degree of dependence on foreign trade? How far do the inducements operating under competition, or State or monopolistic direction, or through systems of consultation lead people to the decisions they really desire?

The pattern of consumption and savings is determined for Ricardo—translating once again into modern terminology—by at least five factors.

(1) The marginal utility of any article diminishes the more one has of it. Ricardo is not always consistent about this. There are passages in which he seems to treat a double quantity of a certain article or service as having for its owner twice the utility of a single quantity.

(2) Patterns of consumption change as incomes change.

(3) Social groups or classes have their own solidarity and customs. We would say to-day that preferences are inter-dependent.

(4) Social stability or instability also affect consumption and savings. The steady-going, relatively secure English labourer is contrasted with the 'vicissitudes and miseries' of Ireland or Poland.

(5) The social structure, norms of conduct, and patterns of personality which underlie the previous factors may be changing, or perhaps may not be changing enough. The problem of an under-developed country is as likely as not to be the 'ignorance, indolence, and barbarism of the inhabitants,' or their 'bad government . . . insecurity of property,' and '. . . want of education in all ranks of the people.'

One would wish to-day to group together what Ricardo says in scattered passages on these matters, and to phrase it more exactly. Often it would be possible now to give more statistical precision to his ideas, as for instance to the idea of consumption patterns changing as incomes rise. And an economist to-day would pay more attention than Ricardo does to the effect on consumption and saving patterns of the big indivisible items of expenditure round which a budget may have to be built; such things as the purchase and upkeep of a house or a car. But by and large it would be a case of gilding Ricardo's lily, not of adding much that is new. His argument here is thin and scattered over many separate passages of the *Principles*, but is basically sound.

In the case of investment, Ricardo relies once again on the forces of competition. Under full or perfect competition, he argues, savings will be invested in those activities where the value of the marginal product of capital is highest. The value of the marginal product of capital will not fall in any country merely because of an increase in the proportion of capital to population, for the supply of labour is highly elastic. Let the demand for labour increase and so also will the birth-rate. The accumulation of workers will thus keep pace with that of capital, and the value of the marginal product of neither will change. If this were all that had to be considered, an unlimited amount of capital could be invested without the inducement to invest diminishing. But, Ricardo adds, workers eat, and food is often produced under conditions of rising marginal cost. Where this is so—unless, that is, there is unlimited good land available, or a succession of advances take place in farming technique—wages will have to rise as the number of workers increases. For workers have a customary standard of living, and 'it is impossible to conceive that the money price of wages should fall or remain stationary with a gradually increasing price of necessaries.' A development will then take place resembling—though it does not happen for the same reasons— that which modern economists describe in the theory of imperfect competition. To discover the net return on an investment, there must be deducted from the gross value of the marginal product of capital the increase in the wage bill per unit of capital arising from the increase in the number of workers called forth by the marginal investment, and from the consequent increase in the rate of wages paid to all workers. As investment grows the number of workers increases, food prices and therefore wages rise higher and higher, and the net marginal return on capital becomes smaller and smaller. Investment is discouraged, and eventually ceases. As investment drops, so also do savings. For no one earns an income except to invest or consume. What he does not invest he spends on consumption goods.

There may of course, Ricardo agrees, be temporary divergences from these general trends. Population, for instance, may take a little time to catch up with an accumulation of capital. And there may be snags in the transition from one pattern of transactions to another. A sudden change such as occurs at the end of a war-time boom (Ricardo's example) may cause temporary dislocation. Or mechanization may displace labour, and it will depend on how those who benefit from lower prices or higher profits use their increased incomes—whether, in particular, they do or do not quickly save out of them and create new openings for employment—how soon, if at all, the displaced workers can be reabsorbed.

This is in many ways the least satisfactory part of the *Principles*. The whole argument is confused, and even as regards investment under full or perfect competition Ricardo is in several ways demonstrably wrong. It is true that the growth of population was keeping pace with the growth of capital in the particular case of England in and around his lifetime. But there is no reason to suppose that it will do so in all cases, as indeed Ricardo himself recognizes in other connections: and in fact it often does not. The value of the marginal product of capital can, therefore, and often does fall even when there is no question of increased food costs forcing up labour's share of the national income. And, on the other hand, it does not follow that if food is being produced under conditions of rising marginal cost an increase in the demand for food, and therefore in its cost and price, will lead under effective or perfect competition to a corresponding rise in money wages and fall in the net marginal product of capital. That would follow only if the number of workers increased less than in proportion to the increase in capital and the value of the marginal product of those remaining in the market therefore rose till it equalled the new rate of wages. This can and does happen in practice, but it is not what Ricardo assumes. In his analysis, the number of workers *is* assumed to increase in proportion to the increase in capital. Where that assumption is fulfilled, it is perfectly possible, in the absence of state or union control, for competition to force their wages down to or even below the customary subsistence level. Furthermore, when (for whatever reason) the value of the marginal product of capital *does* fall, and investment is discouraged, it does not in the least follow that savings will fall off correspondingly, or that funds diverted from investment will be used instead for consumption. As Keynes pointed out, there is an alternative to both investment and consumption, namely liquidity. Savings not invested can be simply immobilized. The consequence will often be cyclical slumps and chronic under-employment: problems much more formidable and persistent than those temporary snags which Ricardo has in mind.

And all this is without prejudice to the further criticism that, as usual, Ricardo discusses conditions under competition, but has no satisfactory discussion either of the case where competition in the capital market is less than perfect, or of the cases where it would not in any case be expected to cause the right investments to be made. Nor, therefore, does he discuss the machinery by which investment should in such cases as these be guided.

Ricardo's ideas on the economic factors in population policy are sounder than his incidental references when discussing investment would suggest. He insists too heavily, it is true, on the high elasticity of population growth. A little encouragement by rising wages, or a little lack of severity under the Poor Laws, seems to him likely to call forth a great spate of 'improvident marriages' and their offspring. But he is also clear that in favourable circumstances, which means above all in cases where the elasticity of supply of food is high, savings and investment can proceed fast enough to keep the national income rising ahead of population. This may be helped, he adds, if workers can be persuaded to acquire a taste for new products and a higher standard of living, and if the public can be persuaded to disapprove of those who marry improvidently. In these days he would have talked about birth control: but there was too little of that—in its modern sense—in his time for him to bother with. He makes it clear that he is arguing purely on economic grounds, that is on the assumption that higher income equals higher welfare. He leaves it to others to evaluate the non-economic arguments in favour of increasing the population even at some cost in economic advantage. In all this area, modern statistics permit a much clearer view than was open either to Ricardo or to his contemporary Malthus. We can see to-day why and in what way the flood of population which seemed to be welling up in England in and just after their time was exceptional. In other ways also Ricardo's arguments have been developed and refined. But the framework of his views on population was sounder than some of his less cautious references would suggest.

There is one last question on objectives. How much dependence on international trade is desirable? And what of the consequent problems of currency and exchange? Ricardo's position here is very like that on economic control. He sets out well and convincingly the advantages of free multilateral international trade and of exchanges fixed by a rigid gold standard. He sets down the theory of comparative costs, and describes how payments are balanced under free exchange rates, and how prices, profits, and wage rates adjust themselves to international trade movements under a gold standard. He shows how the value of money depends on the quantity of it in circulation: though he does not make enough of possible variations in the velocity of circulation. But, as before, he does not adequately discuss that range of problems of international trade and exchange which competition solves only partially, if at all, and which has increasingly preoccupied both economists and practitioners in more recent times. International cartels, the unequal international distribution of income and wealth, the international impact of the trade cycle, the stabilization of primary product markets, the need for international research, statistical, and planning services, the need to underpin the economic basis of countries or regions or to revolutionize the economic and social structure of under-developed areas—these problems enter into Ricardo's thought chiefly, when at all, to the extent that they are or might be solved through competition. The modern world has learnt that they cannot be solved without using non-competitive mechanisms as well: which raises once again the question of which mechanism can best be used to solve each problem in each set of circumstances, and of how a code of behaviour can be set up and policed to encourage the use of approved mechanisms and discourage the rest. But only the eye of extreme faith can read much thinking of that sort into Ricardo's *Principles*.

(3) *The scope and method of economics*.

What, finally, is the purpose and method of economic science? For Ricardo the aim of economic activity is to increase what he called the riches of the country, and we to-day would call the real national income. The most ticklish problem of economic science, as he saw it, is to decide how this income is to be shared among the different factors of production; who is to get what, and by which mechanisms. And this problem is to be sorted out in strictly scientific fashion by posing clear-cut hypotheses, such as that of perfect competition, and reasoning from them to see to what conclusions each would lead. The hypotheses are of course suggested by, and are also to be checked against, conditions observed in economic practice.

We to-day can accept this approach: only we state its implications more precisely, and work them out with better tools. Welfare economics has delved deeper than Ricardo could into the question of what constitutes 'economic welfare,' and of the relation this bears to 'welfare' without qualification. The other social sciences have caught up on economics and come of age: politics, anthropology, sociology, psychology, and such cross-cut studies as demography, public administration, or industrial relations. The economist's role can as a result be more precisely defined to-day than in the nineteenth century, and he can learn from other social scientists and they from him. And the economist's statistical tools have been vastly improved. Accountancy has linked up with theoretical economics, bringing to birth in the process social accounting and econometrics. Ricardo spent much time considering the question—one should rather say the hypothesis—of an 'invariable' standard of value. He did not find one. Nor have modern economists. But with the help of index numbers they have come a great deal closer to it than he could ever have done.

One lays down this book with the feeling of having been in contact with a genuinely great man. Awkward and graceless Ricardo, as a writer, no doubt was. His terminology is to-day out of date. He saw clearly only in the limited field of competition. But whoever breaks the language barrier and gets to know him realizes that, in his own special field, he saw for his time very clearly indeed. The foundations he laid were good, and others could build on them. We in more recent times have restated his theory of perfect or effective competition in new language, and have branched out from it into new fields. But we have not essentially altered it. And though we have acquired new tools, and new insights into economics, we still practice essentially the same scientific method which Ricardo developed in the early days; dry, perhaps, and often repellent, but the one sure foundation.

MICHAEL P. FOGARTY.

SELECT BIBLIOGRAPHY

The Works and Correspondence of David Ricardo (ed. Sraffa), Cambridge, 1951: 9 vols., and a volume of biographical material. This is a complete and definitive edition. Ricardo's main works include:

The High Price of Bullion, 1810; A Reply to Mr Bosanquet's Practical Observations on the Report of the Bullion Committee, 1811; Essay on the Influence of a Low Price of Corn on the Profits of Stock, 1815; Proposals for an Economical and Secure Currency, 1816; 'Essay on the Funding System' (in Supplement to the Encyclopaedia Britannica), 1820; Principles of Political Economy and Taxation, 1817; 2nd edition, 1819; 3rd edition, 1821; On Protection to Agriculture, 1822; Plan for the Establishment of a National Bank, 1824.

See N. S. Patten, Malthus and Ricardo, 1889; J. H. Hollander, David Ricardo, 1910; M. Blaug, Ricardian Economics (Yale), 1958; C. S. Shoup, Ricardo on Taxation (New York), 1960; E. C. K. Gonner, The Economic Essays of David Ricardo, 1966.

CONTENTS

		PAGE
INTRODUCTION by Michael P. Fogarty		v
ORIGINAL PREFACE		I
ADVERTISEMENT		3

CHAP.

I.	ON VALUE	5
II.	ON RENT	33
III.	ON THE RENT OF MINES	46
IV.	ON NATURAL AND MARKET PRICE	48
V.	ON WAGES	52
VI.	ON PROFITS	64
VII.	ON FOREIGN TRADE	77
VIII.	ON TAXES	94
IX.	TAXES ON RAW PRODUCE	98
X.	TAXES ON RENT	110
XI.	TITHES	112
XII.	LAND-TAX	115
XIII.	TAXES ON GOLD	122
XIV.	TAXES ON HOUSES	129
XV.	TAXES ON PROFITS	132
XVI.	TAXES ON WAGES	140
XVII.	TAXES ON OTHER COMMODITIES THAN RAW PRODUCE .	160
XVIII.	POOR RATES	171
XIX.	ON SUDDEN CHANGES IN THE CHANNELS OF TRADE .	175
XX.	VALUE AND RICHES, THEIR DISTINCTIVE PROPERTIES .	182
XXI.	EFFECTS OF ACCUMULATION ON PROFITS AND INTEREST	192
XXII.	BOUNTIES ON EXPORTATION, AND PROHIBITIONS OF IMPORTATION	201
XXIII.	ON BOUNTIES ON PRODUCTION	215
XXIV.	DOCTRINE OF ADAM SMITH CONCERNING THE RENT OF LAND	219

xvii

CHAP.		PAGE
XXV.	On Colonial Trade	227
XXVI.	On Gross and Net Revenue	234
XXVII.	On Currency and Banks	238
XXVIII.	On the Comparative Value of Gold, Corn, and Labour, in Rich and Poor Countries . .	253
XXIX.	Taxes Paid by the Producer	258
XXX.	On the Influence of Demand and Supply on Prices .	260
XXXI.	On Machinery	263
XXXII.	Mr. Malthus's Opinions on Rent	272
	Index	293

ORIGINAL PREFACE

THE produce of the earth—all that is derived from its surface by the united application of labour, machinery, and capital, is divided among three classes of the community, namely, the proprietor of the land, the owner of the stock or capital necessary for its cultivation, and the labourers by whose industry it is cultivated.

But in different stages of society, the proportions of the whole produce of the earth which will be allotted to each of these classes, under the names of rent, profit, and wages, will be essentially different; depending mainly on the actual fertility of the soil, on the accumulation of capital and population, and on the skill, ingenuity, and instruments employed in agriculture.

To determine the laws which regulate this distribution is the principal problem in Political Economy: much as the science has been improved by the writings of Turgot, Stuart, Smith, Say, Sismondi, and others, they afford very little satisfactory information respecting the natural course of rent, profit, and wages.

In 1815, Mr. Malthus, in his *Inquiry into the Nature and Progress of Rent*, and a Fellow of University College, Oxford, in his *Essay on the Application of Capital to Land*, presented to the world, nearly at the same moment, the true doctrine of rent; without a knowledge of which it is impossible to understand the effect of the progress of wealth on profits and wages, or to trace satisfactorily the influence of taxation on different classes of the community; particularly when the commodities taxed are the productions immediately derived from the surface of the earth. Adam Smith, and the other able writers to whom I have alluded, not having viewed correctly the principles of rent, have, it appears to me, overlooked many important truths, which can only be discovered after the subject of rent is thoroughly understood.

To supply this deficiency, abilities are required of a far superior cast to any possessed by the writer of the following pages; yet, after having given to this subject his best consideration—after the aid which he has derived from the works of the above-mentioned eminent writers—and after the valuable experience which a few late years, abounding in facts, have yielded to the present generation—it will not, he trusts, be deemed presumptuous in him to state his opinions on the laws of profits and wages, and on the operation of taxes. If the principles which he deems correct should be found to be so, it will be for others, more able than himself, to trace them to all their important consequences.

The writer, in combating received opinions, has found it necessary to advert more particularly to those passages in the writings of Adam Smith from which he sees reason to differ; but he hopes it will not, on that account, be suspected that he does not, in common with all those who acknowledge the importance of the science of Political Economy, participate in the admiration which the profound work of this celebrated author so justly excites.

The same remark may be applied to the excellent works of M. Say, who not only was the first, or among the first, of continental writers who justly appreciated and applied the principles of Smith, and who has done more than all other continental writers taken together to recommend the principles of that enlightened and beneficial system to the nations of Europe; but who has succeeded in placing the science in a more logical and more instructive order; and has enriched it by several discussions, original, accurate, and profound.[1] The respect, however, which the author entertains for the writings of this gentleman has not prevented him from commenting with that freedom which he thinks the interests of science require, on such passages of the *Economie Politique* as appeared at variance with his own ideas.

[1] Chap. xv. Part i., *Des Débouchés*, contains, in particular, some very important principles, which I believe were first explained by this distinguished writer.

ADVERTISEMENT TO THE THIRD EDITION

In this edition I have endeavoured to explain more fully than in the last my opinion on the difficult subject of Value, and for that purpose have made a few additions to the first chapter. I have also inserted a new chapter on the subject of Machinery, and on the effects of its improvement on the interests of the different classes of the state. In the chapter on the Distinctive Properties of Value and Riches, I have examined the doctrines of M. Say on that important question, as amended in the fourth and last edition of his work. I have in the last chapter endeavoured to place in a stronger point of view than before the doctrine of the ability of a country to pay additional money taxes, although the aggregate money value of the mass of its commodities should fall, in consequence either of the diminished quantity of labour required to produce its corn at home, by improvements in its husbandry, or from its obtaining a part of its corn at a cheaper price from abroad, by means of the exportation of its manufactured commodities. This considera-tion is of great importance, as it regards the question of the policy of leaving unrestricted the importation of foreign corn, particularly in a country burthened with a heavy fixed money taxation, the consequence of an immense National Debt. I have endeavoured to show that the ability to pay taxes depends, not on the gross money value of the mass of commodities, nor on the net money value of the revenues of capitalists and land-lords, but on the money value of each man's revenue compared to the money value of the commodities which he usually consumes.

March 26, 1821.

CHAPTER I

SECTION I

The value of a commodity, or the quantity of any other commodity for which it will exchange, depends on the relative quantity of labour which is necessary for its production, and not on the greater or less compensation which is paid for that labour

It has been observed by Adam Smith that " the word Value has two different meanings, and sometimes expresses the utility of some particular object, and sometimes the power of purchasing other goods which the possession of that object conveys. The one may be called *value in use ;* the other *value in exchange.* The things," he continues, " which have the greatest value in use, have frequently little or no value in exchange; and, on the contrary, those which have the greatest value in exchange, have little or no value in use." Water and air are abundantly useful; they are indeed indispensable to existence, yet, under ordinary circumstances, nothing can be obtained in exchange for them. Gold, on the contrary, though of little use compared with air or water, will exchange for a great quantity of other goods.

Utility then is not the measure of exchangeable value, although it is absolutely essential to it. If a commodity were in no way useful—in other words, if it could in no way contribute to our gratification—it would be destitute of exchangeable value, however scarce it might be, or whatever quantity of labour might be necessary to procure it.

Possessing utility, commodities derive their exchangeable value from two sources: from their scarcity, and from the quantity of labour required to obtain them.

5

There are some commodities, the value of which is determined by their scarcity alone. No labour can increase the quantity of such goods, and therefore their value cannot be lowered by an increased supply. Some rare statues and pictures, scarce books and coins, wines of a peculiar quality, which can be made only from grapes grown on a particular soil, of which there is a very limited quantity, are all of this description. Their value is wholly independent of the quantity of labour originally necessary to produce them, and varies with the varying wealth and inclinations of those who are desirous to possess them.

These commodities, however, form a very small part of the mass of commodities daily exchanged in the market. By far the greatest part of those goods which are the objects of desire are procured by labour; and they may be multiplied, not in one country alone, but in many, almost without any assignable limit, if we are disposed to bestow the labour necessary to obtain them.

In speaking, then, of commodities, of their exchangeable value, and of the laws which regulate their relative prices, we mean always such commodities only as can be increased in quantity by the exertion of human industry, and on the production of which competition operates without restraint.

In the early stages of society, the exchangeable value of these commodities, or the rule which determines how much of one shall be given in exchange for another, depends almost exclusively on the comparative quantity of labour expended on each.

" The real price of everything," says Adam Smith, " what everything really costs to the man who wants to acquire it, is the toil and trouble of acquiring it. What everything is really worth to the man who has acquired it, and who wants to dispose of it, or exchange it for something else, is the toil and trouble which it can save to himself, and which it can impose upon other people." " Labour was the first price—the original purchase-money that was paid for all things." Again, " in that early and rude state of society which precedes both the accumulation of stock and the appropriation of land, the proportion between the quantities of labour necessary for acquiring different objects seems to be the only circumstance which can afford any rule for exchanging them for one another. If, among a nation of hunters, for example, it usually cost twice the labour to kill a beaver which it does to kill a deer, one beaver should naturally exchange for, or be worth, two deer. It is natural that what is usually the produce of two days' or two hours' labour should be worth double of what is usually the produce of one day's or one hour's labour." [1]

That this is really the foundation of the exchangeable value of all things, excepting those which cannot be increased by human industry, is a doctrine of the utmost importance in political economy; for from no source do so many errors, and so much difference of opinion in that science proceed, as from the vague ideas which are attached to the word value.

If the quantity of labour realised in commodities regulate their exchangeable value, every increase of the quantity of labour must augment the value of that commodity on which it is exercised, as every diminution must lower it.

Adam Smith, who so accurately defined the original source of exchangeable value, and who was bound in consistency to maintain that all things became more or less valuable in proportion as more or less labour was bestowed on their production, has himself erected another standard measure of value, and speaks of things being more or less valuable in proportion as they will exchange for more or less of this standard measure. Sometimes he speaks of corn, at other times of labour, as a standard measure; not the quantity of labour bestowed on the production of any object, but the quantity which it can command in the market: as if these were two equivalent expressions, and as if, because a man's labour had become doubly efficient, and he could therefore produce twice the quantity of a commodity, he would necessarily receive twice the former quantity in exchange for it.

If this indeed were true, if the reward of the labourer were always in proportion to what he produced, the quantity of labour bestowed on a commodity, and the quantity of labour which that commodity would purchase, would be equal, and either might accurately measure the variations of other things; but they are not equal; the first is under many circumstances an invariable standard, indicating correctly the variations of other things; the latter is subject to as many fluctuations as the commodities compared with it. Adam Smith, after most ably showing the insufficiency of a variable medium, such as gold and silver, for the purpose of determining the varying value of other things, has himself, by fixing on corn or labour, chosen a medium no less variable.

Gold and silver are no doubt subject to fluctuations from the discovery of new and more abundant mines; but such discoveries are rare, and their effects, though powerful, are limited

[1] Book i. chap. 5.

to periods of comparatively short duration. They are subject also to fluctuation from improvements in the skill and machinery with which the mines may be worked; as in consequence of such improvements a greater quantity may be obtained with the same labour. They are further subject to fluctuation from the decreasing produce of the mines, after they have yielded a supply to the world for a succession of ages. But from which of these sources of fluctuation is corn exempted? Does not that also vary, on one hand, from improvements in agriculture, from improved machinery and implements used in husbandry, as well as from the discovery of new tracts of fertile land, which in other countries may be taken into cultivation, and which will affect the value of corn in every market where importation is free? Is it not on the other hand subject to be enhanced in value from prohibitions of importation, from increasing population and wealth, and the greater difficulty of obtaining the increased supplies, on account of the additional quantity of labour which the cultivation of inferior land requires? Is not the value of labour equally variable; being not only affected, as all other things are, by the proportion between the supply and demand, which uniformly varies with every change in the condition of the community, but also by the varying price of food and other necessaries, on which the wages of labour are expended?

In the same country double the quantity of labour may be required to produce a given quantity of food and necessaries at one time that may be necessary at another and a distant time; yet the labourer's reward may possibly be very little diminished. If the labourer's wages at the former period were a certain quantity of food and necessaries, he probably could not have subsisted if that quantity had been reduced. Food and necessaries in this case will have risen 100 per cent. if estimated by the *quantity* of labour necessary to their production, while they will scarcely have increased in value if measured by the quantity of labour for which they will *exchange*.

The same remark may be made respecting two or more countries. In America and Poland, on the land last taken into cultivation, a year's labour of any given number of men will produce much more corn than on land similarly circumstanced in England. Now, supposing all other necessaries to be equally cheap in those three countries, would it not be a great mistake to conclude that the quantity of corn awarded to the labourer would in each country be in proportion to the facility of production?

If the shoes and clothing of the labourer could, by improvements in machinery, be produced by one-fourth of the labour now necessary to their production, they would probably fall 75 per cent.; but so far is it from being true that the labourer would thereby be enabled permanently to consume four coats, or four pair of shoes, instead of one, that it is probable his wages would in no long time be adjusted by the effects of competition, and the stimulus to population, to the new value of the necessaries on which they were expended. If these improvements extended to all the objects of the labourer's consumption, we should find him probably, at the end of a very few years, in possession of only a small, if any, addition to his enjoyments, although the exchangeable value of those commodities, compared with any other commodity, in the manufacture of which no such improvement were made, had sustained a very considerable reduction; and though they were the produce of a very considerably diminished quantity of labour.

It cannot then be correct to say with Adam Smith, " that as labour may sometimes *purchase* a greater and sometimes a smaller quantity of goods, it is their value which varies, not that of the labour which purchases them; " and therefore, " that labour, *alone never varying in its own value*, is alone the ultimate and real standard by which the value of all commodities can at all times and places be estimated and compared; " —but it is correct to say, as Adam Smith had previously said, " that the proportion between the quantities of labour necessary for acquiring different objects seems to be the only circumstance which can afford any rule for exchanging them for one another; " or in other words that it is the comparative quantity of commodities which labour will produce that determines their present or past relative value, and not the comparative quantities of commodities which are given to the labourer in exchange for his labour.

Two commodities vary in relative value, and we wish to know in which the variation has really taken place. If we compare the present value of one with shoes, stockings, hats, iron, sugar, and all other commodities, we find that it will exchange for precisely the same quantity of all these things as before. If we compare the other with the same commodities, we find it has varied with respect to them all: we may then with great probability infer that the variation has been in this commodity, and not in the commodities with which we have compared it. If on examining still more particularly into all the circumstances connected with the production of these various commodities, we find that precisely the same quantity of labour and capital are necessary to the production of the shoes, stockings, hats, iron, sugar, etc.; but that the same quantity as before is not necessary to produce the single commodity whose relative value is altered, probability is changed into certainty, and we are sure that the variation is in the single commodity: we then discover also the cause of its variation.

If I found that an ounce of gold would exchange for a less quantity of all the commodities above enumerated and many others; and if, moreover, I found that by the discovery of a new and more fertile mine, or by the employment of machinery to great advantage, a given quantity of gold could be obtained with a less quantity of labour, I should be justified in saying that the cause of the alteration in the value of gold relatively to other commodities was the greater facility of its production, or the smaller quantity of labour necessary to obtain it. In like manner, if labour fell very considerably in value, relatively to all other things, and if I found that its fall was in consequence of an abundant supply, encouraged by the great facility with which corn, and the other necessaries of the labourer, were produced, it would, I apprehend, be correct for me to say that corn and necessaries had fallen in value in consequence of less quantity of labour being necessary to produce them, and that this facility of providing for the support of the labourer had been followed by a fall in the value of labour. No, say Adam Smith and Mr. Malthus, in the case of the gold you were correct in calling its variation a fall of its value, because corn and labour had not then varied; and as gold would command a less quantity of them, as well as of all other things, than before, it was correct to say that all things had remained stationary and that gold only had varied; but when corn and labour fall, things which we have selected to be our standard measure of value, notwithstanding all the variations to which we acknowledge they are subject, it would be highly improper to say so; the correct language will be to say that corn and labour have remained stationary, and all other things have risen in value.

Now it is against this language that I protest. I find that precisely, as in the case of the gold, the cause of the variation between corn and other things is the smaller quantity of labour necessary to produce it, and therefore, by all just reasoning, I am bound to call the variation of corn and labour a fall in their value, and not a rise in the value of the things with which they are compared. If I have to hire a labourer for a week, and instead of ten shillings I pay him eight, no variation having taken place in the value of money, the labourer can probably obtain more food and necessaries with his eight shillings than he before obtained for ten: but this is owing, not to a rise in the real value of his wages, as stated by Adam Smith, and more recently by Mr. Malthus, but to a fall in the value of the things on which his wages are expended, things perfectly distinct; and yet for calling this a fall in the real value of wages, I am told that I adopt new and unusual language, not reconcilable with the true principles of the science. To me it appears that the unusual and, indeed, inconsistent language is that used by my opponents.

Suppose a labourer to be paid a bushel of corn for a week's work when the price of corn is 80s. per quarter, and that he is paid a bushel and a quarter when the price falls to 40s. Suppose, too, that he consumes half a bushel of corn a week in his own family, and exchanges the remainder for other things, such as fuel, soap, candles, tea, sugar, salt, etc. etc.; if the three-fourths of a bushel which will remain to him, in one case, cannot procure him as much of the above commodities as half a bushel did in the other, which it will not, will labour have risen or fallen in value? Risen, Adam Smith must say, because his standard is corn, and the labourer receives more corn for a week's labour. Fallen, must the same Adam Smith say, "because the value of a thing depends on the power of purchasing other goods which the possession of that object conveys," and labour has a less power of purchasing such other goods.

SECTION II

Labour of different qualities differently rewarded. This no cause of variation in the relative value of commodities

In speaking, however, of labour, as being the foundation of all value, and the relative quantity of labour as almost exclusively determining the relative value of commodities, I must not be supposed to be inattentive to the different qualities of labour, and the difficulty of comparing an hour's or a day's labour in one employment with the same duration of labour in another. The estimation in which different qualities of labour are held comes soon to be adjusted in the market with sufficient precision for all practical purposes, and depends much on the comparative skill of the labourer and intensity of the labour performed. The scale, when once formed, is liable to little variation. If a day's labour of a working jeweller be more valuable than a day's labour of a common labourer, it has long ago been adjusted and placed in its proper position in the scale of value.[1]

In comparing, therefore, the value of the same commodity at different periods of time, the consideration of the comparative skill and intensity of labour required for that particular commodity needs scarcely to be attended to, as it operates equally at both periods. One description of labour at one time is compared with the same description of labour at another; if a tenth, a fifth, or a fourth has been added or taken away, an effect proportioned to the cause will be produced on the relative value of the commodity.

If a piece of cloth be now of the value of two pieces of linen, and if, in ten years hence, the ordinary value of a piece of cloth should be four pieces of linen, we may safely conclude that either more labour is required to make the cloth, or less to make the linen, or that both causes have operated.

As the inquiry to which I wish to draw the reader's attention relates to the effect of the variations in the relative value of commodities, and not in their absolute value, it will be of little importance to examine into the comparative degree of estimation in which the different kinds of human labour are held. We may fairly conclude that whatever inequality there might originally have been in them, whatever the ingenuity, skill, or time necessary for the acquirement of one species of manual dexterity more than another, it continues nearly the same from one generation to another; or at least that the variation is very inconsiderable from year to year, and therefore can

---

[1] " But though labour be the real measure of the exchangeable value of all commodities, it is not that by which their value is commonly estimated. It is often difficult to ascertain the proportion between two different quantities of labour. The time spent in two different sorts of work will not always alone determine this proportion. The different degrees of hardship endured, and of ingenuity exercised, must likewise be taken into account. There may be more labour in an hour's hard work than in two hours' easy business; or in an hour's application to a trade, which it costs ten years' labour to learn, than in a month's industry at an ordinary and obvious employment. But it is not easy to find any accurate measure, either of hardship or ingenuity. In exchanging, indeed, the different productions of different sorts of labour for one another, some allowance is commonly made for both. It is adjusted, however, not by any accurate measure, but by the higgling and bargaining of the market, according to that sort of rough equality which, though not exact, is sufficient for carrying on the business of common life."—*Wealth of Nations*, book i. chap. 10.

have little effect, for short periods, on the relative value of commodities.

"The proportion between the different rates both of wages and profit in the different employments of labour and stock seems not to be much affected, as has already been observed, by the riches or poverty, the advancing, stationary, or declining state of the society. Such revolutions in the public welfare, though they affect the general rates both of wages and profit, must in the end affect them equally in all different employments. The proportion between them therefore must remain the same, and cannot well be altered, at least for any considerable time, by any such revolutions." [1]

SECTION III

Not only the labour applied immediately to commodities affect their value, but the labour also which is bestowed on the implements, tools, and buildings, with which such labour is assisted

EVEN in that early state to which Adam Smith refers, some capital, though possibly made and accumulated by the hunter himself, would be necessary to enable him to kill his game. Without some weapon, neither the beaver nor the deer could be destroyed, and therefore the value of these animals would be regulated, not solely by the time and labour necessary to their destruction, but also by the time and labour necessary for providing the hunter's capital, the weapon, by the aid of which their destruction was effected.

Suppose the weapon necessary to kill the beaver was constructed with much more labour than that necessary to kill the deer, on account of the greater difficulty of approaching near to the former animal, and the consequent necessity of its being more true to its mark; one beaver would naturally be of more value than two deer, and precisely for this reason, that more labour would, on the whole, be necessary to its destruction. Or suppose that the same quantity of labour was necessary to make both weapons, but that they were of very unequal durability; of the durable implement only a small portion of its value would be transferred to the commodity, a much greater portion of the value of the less durable implement would be realised in the commodity which it contributed to produce.

All the implements necessary to kill the beaver and deer might belong to one class of men, and the labour employed in

[1] *Wealth of Nations*, book i. chap. 10.

their destruction might be furnished by another class; still, their comparative prices would be in proportion to the actual labour bestowed, both on the formation of the capital and on the destruction of the animals. Under different circumstances of plenty or scarcity of capital, as compared with labour, under different circumstances of plenty or scarcity of the food and necessaries essential to the support of men, those who furnished an equal value of capital for either one employment or for the other might have a half, a fourth, or an eighth of the produce obtained, the remainder being paid as wages to those who furnished the labour; yet this division could not affect the relative value of these commodities, since whether the profits of capital were greater or less, whether they were 50, 20, or 10 per cent., or whether the wages of labour were high or low, they would operate equally on both employments.

If we suppose the occupations of the society extended, that some provide canoes and tackle necessary for fishing, others the seed and rude machinery first used in agriculture, still the same principle would hold true, that the exchangeable value of the commodities produced would be in proportion to the labour bestowed on their production; not on their immediate production only, but on all those implements or machines required to give effect to the particular labour to which they were applied.

If we look to a state of society in which greater improvements have been made, and in which arts and commerce flourish, we shall still find that commodities vary in value conformably with this principle: in estimating the exchangeable value of stockings, for example, we shall find that their value, comparatively with other things, depends on the total quantity of labour necessary to manufacture them and bring them to market. First, there is the labour necessary to cultivate the land on which the raw cotton is grown; secondly, the labour of conveying the cotton to the country where the stockings are to be manufactured, which includes a portion of the labour bestowed in building the ship in which it is conveyed, and which is charged in the freight of the goods; thirdly, the labour of the spinner and weaver; fourthly, a portion of the labour of the engineer, smith, and carpenter, who erected the buildings and machinery, by the help of which they are made; fifthly, the labour of the retail dealer, and of many others, whom it is unnecessary further to particularise. The aggregate sum of these various kinds of labour determines the quantity of other things for which these stockings will exchange, while the same consideration of the various quantities of labour which have been bestowed on those other things will equally govern the portion of them which will be given for the stockings.

To convince ourselves that this is the real foundation of exchangeable value, let us suppose any improvement to be made in the means of abridging labour in any one of the various processes through which the raw cotton must pass before the manufactured stockings come to the market to be exchanged for other things, and observe the effects which will follow. If fewer men were required to cultivate the raw cotton, or if fewer sailors were employed in navigating, or shipwrights in constructing the ship, in which it was conveyed to us; if fewer hands were employed in raising the buildings and machinery, or if these, when raised, were rendered more efficient, the stockings would inevitably fall in value, and consequently command less of other things. They would fall, because a less quantity of labour was necessary to their production, and would therefore exchange for a smaller quantity of those things in which no such abridgment of labour had been made.

Economy in the use of labour never fails to reduce the relative value of a commodity, whether the saving be in the labour necessary to the manufacture of the commodity itself, or in that necessary to the formation of the capital by the aid of which it is produced. In either case the price of stockings would fall, whether there were fewer men employed as bleachers, spinners, and weavers, persons immediately necessary to their manufacture; or as sailors, carriers, engineers, and smiths, persons more indirectly concerned. In the one case, the whole saving of labour would fall on the stockings, because that portion of labour was wholly confined to the stockings; in the other, a portion only would fall on the stockings, the remainder being applied to all those other commodities, to the production of which the buildings, machinery, and carriage were subservient.

Suppose that, in the early stages of society, the bows and arrows of the hunter were of equal value, and of equal durability, with the canoe and implements of the fisherman, both being the produce of the same quantity of labour. Under such circumstances the value of the deer, the produce of the hunter's day's labour, would be exactly equal to the value of the fish, the produce of the fisherman's day's labour. The comparative value of the fish and the game would be entirely regulated by the quantity of labour realised in each, whatever might be the quantity of production or however high or low general wages or profits might be. If, for example, the canoes and implements of the fisherman were of the value of £100, and were calculated to last for ten years, and he employed ten men, whose annual labour cost £100, and who in one day obtained by their labour twenty salmon: If the weapons employed by the hunter were also of £100 value, and calculated to last ten years, and if he also employed ten men, whose annual labour cost £100, and who in one day procured him ten deer; then the natural price of a deer would be two salmon, whether the proportion of the whole produce bestowed on the men who obtained it were large or small. The proportion which might be paid for wages is of the utmost importance in the question of profits; for it must at once be seen that profits would be high or low exactly in proportion as wages were low or high; but it could not in the least affect the relative value of fish and game, as wages would be high or low at the same time in both occupations. If the hunter urged the plea of his paying a large proportion, or the value of a large proportion of his game for wages, as an inducement to the fisherman to give him more fish in exchange for his game, the latter would state that he was equally affected by the same cause; and therefore, under all variations of wages and profits, under all the effects of accumulation of capital, as long as they continued by a day's labour to obtain respectively the same quantity of fish and the same quantity of game, the natural rate of exchange would be one deer for two salmon.

If with the same quantity of labour a less quantity of fish or a greater quantity of game were obtained, the value of fish would rise in comparison with that of game. If, on the contrary, with the same quantity of labour a less quantity of game or a greater quantity of fish was obtained, game would rise in comparison with fish.

If there were any other commodity which was invariable in its value, we should be able to ascertain, by comparing the value of fish and game with this commodity, how much of the variation was to be attributed to a cause which affected the value of fish, and how much to a cause which affected the value of game.

Suppose money to be that commodity. If a salmon were worth £1 and a deer £2, one deer would be worth two salmon. But a deer might become of the value of three salmon, for more labour might be required to obtain the deer, or less to get the salmon, or both these causes might operate at the same time. If we had this invariable standard, we might easily ascertain in what degree either of these causes operated. If salmon continued to sell for £1 whilst deer rose to £3, we might conclude that more labour was required to obtain the deer. If deer continued at the same price of £2 and salmon sold for 13s. 4d., we might then be sure that less labour was required to obtain the salmon; and if deer rose to £2 10s. and salmon fell to 16s. 8d., we should be convinced that both causes had operated in producing the alteration of the relative value of these commodities.

No alteration in the wages of labour could produce any alteration in the relative value of these commodities; for suppose them to rise, no greater quantity of labour would be required in any of these occupations but it would be paid for at a higher price, and the same reasons which should make the hunter and fisherman endeavour to raise the value of their game and fish would cause the owner of the mine to raise the value of his gold. This inducement acting with the same force on all these three occupations, and the relative situation of those engaged in them being the same before and after the rise of wages, the relative value of game, fish, and gold would continue unaltered. Wages might rise twenty per cent., and profits consequently fall in a greater or less proportion, without occasioning the least alteration in the relative value of these commodities.

Now suppose that, with the same labour and fixed capital, more fish could be produced, but no more gold or game, the relative value of fish would fall in comparison with gold or game. If, instead of twenty salmon, twenty-five were the produce of one day's labour, the price of a salmon would be sixteen shillings instead of a pound, and two salmon and a half, instead of two salmon, would be given in exchange for one deer, but the price of deer would continue at £2 as before. In the same manner, if fewer fish could be obtained with the same capital and labour, fish would rise in comparative value. Fish then would rise or fall in exchangeable value, only because more or less labour was required to obtain a given quantity; and it never could rise or fall beyond the proportion of the increased or diminished quantity of labour required.

If we had then an invariable standard, by which we could measure the variation in other commodities, we should find that the utmost limit to which they could permanently rise, if produced under the circumstances supposed, was proportioned to the additional quantity of labour required for their production; and that unless more labour were required for their production they could not rise in any degree whatever. A rise of wages would not raise them in money value, nor relatively to any other commodities, the production of which required no additional quantity of labour, which employed the same proportion of fixed and circulating capital, and fixed capital of the same durability. If more or less labour were required in the production of the other commodity, we have already stated that this will immediately occasion an alteration in its relative value, but such alteration is owing to the altered quantity of requisite labour, and not to the rise of wages.

SECTION IV

The principle that the quantity of labour bestowed on the production of commodities regulates their relative value considerably modified by the employment of machinery and other fixed and durable capital

IN the former section we have supposed the implements and weapons necessary to kill the deer and salmon to be equally durable, and to be the result of the same quantity of labour, and we have seen that the variations in the relative value of deer and salmon depended solely on the varying quantities of labour necessary to obtain them, but in every state of society, the tools, implements, buildings, and machinery employed in different trades may be of various degrees of durability, and may require different portions of labour to produce them. The proportions, too, in which the capital that is to support labour, and the capital that is invested in tools, machinery, and buildings, may be variously combined. This difference in the degree of durability of fixed capital, and this variety in the proportions in which the two sorts of capital may be combined, introduce another cause, besides the greater or less quantity of labour necessary to produce commodities, for the variations in their relative value—this cause is the rise or fall in the value of labour.

The food and clothing consumed by the labourer, the buildings in which he works, the implements with which his labour is assisted, are all of a perishable nature. There is, however, a vast difference in the time for which these different capitals will endure: a steam-engine will last longer than a ship, a ship than the clothing of the labourer, and the clothing of the labourer longer than the food which he consumes.

According as capital is rapidly perishable, and requires to be frequently reproduced, or is of slow consumption, it is classed under the heads of circulating or of fixed capital.[1] A brewer whose buildings and machinery are valuable and durable is said to employ a large portion of fixed capital: on the contrary, a shoemaker, whose capital is chiefly employed in the payment of wages, which are expended on food and clothing, commodities more perishable than buildings and machinery, is said to employ a large proportion of his capital as circulating capital.

It is also to be observed that the circulating capital may circulate, or be returned to its employer, in very unequal times. The wheat bought by a farmer to sow is comparatively a fixed capital to the wheat purchased by a baker to make into loaves. One leaves it in the ground and can obtain no return for a year; the other can get it ground into flour, sell it as bread to his customers, and have his capital free to renew the same or commence any other employment in a week.

Two trades then may employ the same amount of capital; but it may be very differently divided with respect to the portion which is fixed and that which is circulating.

In one trade very little capital may be employed as circulating capital, that is to say, in the support of labour—it may be principally invested in machinery, implements, buildings, etc., capital of a comparatively fixed and durable character. In another trade the same amount of capital may be used, but it may be chiefly employed in the support of labour, and very little may be invested in implements, machines, and buildings. A rise in the wages of labour cannot fail to affect unequally commodities produced under such different circumstances.

Again, two manufacturers may employ the same amount of fixed and the same amount of circulating capital; but the durability of their fixed capitals may be very unequal. One may have steam-engines of the value of £10,000, the other, ships of the same value.

If men employed no machinery in production but labour only, and were all the same length of time before they brought their commodities to market, the exchangeable value of their goods would be precisely in proportion to the quantity of labour employed.

If they employed fixed capital of the same value and of the same durability, then, too, the value of the commodities produced would be the same, and they would vary with the greater or less quantity of labour employed on their production.

[1] A division not essential, and in which the line of demarcation cannot be accurately drawn.

But although commodities produced under similar circumstances would not vary with respect to each other from any
cause but an addition or diminution of the quantity of labour
necessary to produce one or other of them, yet, compared with
others not produced with the same proportionate quantity of
fixed capital, they would vary from the other cause also which
I have before mentioned, namely, a rise in the value of labour,
although neither more nor less labour were employed in the production of either of them.   Barley and oats would continue to
bear the same relation to each other under any variation of wages.
Cotton goods and cloth would do the same, if they also were
produced under circumstances precisely similar to each other,
but yet with a rise or fall of wages barley might be more or less
valuable compared with cotton goods and oats compared with
cloth.

Suppose two men employ one hundred men each for a year
in the construction of two machines, and another man employs
the same number of men in cultivating corn, each of the
machines at the end of the year will be of the same value as the
corn, for they will each be produced by the same quantity of
labour.   Suppose one of the owners of one of the machines to
employ it, with the assistance of one hundred men, the following
year in making cloth, and the owner of the other machine to
employ his also, with the assistance likewise of one hundred
men, in making cotton goods, while the farmer continues to
employ one hundred men as before in the cultivation of corn.
During the second year they will all have employed the same
quantity of labour, but the goods and machine together of the
clothier, and also of the cotton manufacturer, will be the result
of the labour of two hundred men employed for a year; or,
rather, of the labour of one hundred men for two years; whereas
the corn will be produced by the labour of one hundred men
for one year, consequently if the corn be of the value of £500,
the machine and cloth of the clothier together ought to be of
the value of £1000, and the machine and cotton goods of the
cotton manufacturer ought to be also of twice the value of the
corn.   But they will be of more than twice the value of the
corn, for the profit on the clothier's and cotton manufacturer's
capital for the first year has been added to their capitals, while
that of the farmer has been expended and enjoyed.   On account
then of the different degrees of durability of their capitals, or,
which is the same thing, on account of the time which must
elapse before one set of commodities can be brought to market,

they will be valuable, not exactly in proportion to the quantity
of labour bestowed on them—they will not be as two to one,
but something more, to compensate for the greater length of
time which must elapse before the most valuable can be brought
to market.

Suppose that for the labour of each workman £50 per annum
were paid, or that £5000 capital were employed and profits were
10 per cent., the value of each of the machines as well as of the
corn, at the end of the first year, would be £5500. The second
year the manufacturers and farmers will again employ £5000
each in the support of labour, and will therefore again sell their
goods for £5500; but the men using the machines, to be on a
par with the farmer, must not only obtain £5500 for the equal
capitals of £5000 employed on labour, but they must obtain
a further sum of £550 for the profit on £5500, which they have
invested in machinery, and consequently their goods must sell
for £6050. Here, then, are capitalists employing precisely the
same quantity of labour annually on the production of their
commodities, and yet the goods they produce differ in value
on account of the different quantities of fixed capital, or accumulated labour, employed by each respectively. The cloth and
cotton goods are of the same value, because they are the produce
of equal quantities of labour and equal quantities of fixed capital;
but corn is not of the same value as these commodities, because
it is produced, as far as regards fixed capital, under different
circumstances.

But how will their relative value be affected by a rise in the
value of labour? It is evident that the relative values of cloth
and cotton goods will undergo no change, for what affects one
must equally affect the other under the circumstances supposed;
neither will the relative values of wheat and barley undergo
any change, for they are produced under the same circumstances as far as fixed and circulating capital are concerned;
but the relative value of corn to cloth, or to cotton goods, must
be altered by a rise of labour.

There can be no rise in the value of labour without a fall of
profits. If the corn is to be divided between the farmer and the
labourer, the larger the proportion that is given to the latter the
less will remain for the former. So, if cloth or cotton goods be
divided between the workman and his employer, the larger the
proportion given to the former the less remains for the latter.
Suppose then, that owing to a rise of wages, profits fall from
10 to 9 per cent., instead of adding £550 to the common price

of their goods (to £5500) for the profits on their fixed capital, the manufacturers would add only 9 per cent. on that sum, or £495, consequently the price would be £5995 instead of £6050. As the corn would continue to sell for £5500 the manufactured goods in which more fixed capital was employed would fall relatively to corn or to any other goods in which a less portion of fixed capital entered. The degree of alteration in the relative value of goods, on account of a rise or fall of labour, would depend on the proportion which the fixed capital bore to the whole capital employed. All commodities which are produced by very valuable machinery, or in very valuable buildings, or which require a great length of time before they can be brought to market, would fall in relative value, while all those which were chiefly produced by labour, or which would be speedily brought to market, would rise in relative value.

The reader, however, should remark that this cause of the variation of commodities is comparatively slight in its effects. With such a rise of wages as should occasion a fall of 1 per cent. in profits, goods produced under the circumstances I have supposed vary in relative value only 1 per cent.; they fall with so great a fall of profits from £6050 to £5995. The greatest effects which could be produced on the relative prices of these goods from a rise of wages could not exceed 6 or 7 per cent.; for profits could not, probably, under any circumstances, admit of a greater general and permanent depression than to that amount.

Not so with the other great cause of the variation in the value of commodities, namely, the increase or diminution in the quantity of labour necessary to produce them. If to produce the corn, eighty, instead of one hundred men, should be required, the value of the corn would fall 20 per cent., or from £5500 to £4400. If to produce the cloth, the labour of eighty instead of one hundred men would suffice, cloth would fall from £6050 to £4950. An alteration in the permanent rate of profits, to any great amount, is the effect of causes which do not operate but in the course of years, whereas alterations in the quantity of labour necessary to produce commodities are of daily occurrence. Every improvement in machinery, in tools, in buildings, in raising the raw material, saves labour, and enables us to produce the commodity to which the improvement is applied with more facility, and consequently its value alters. In estimating, then, the causes of the variations in the value of commodities, although it would be wrong wholly to omit the consideration of the effect produced by a rise or fall of labour, it would be equally incorrect to attach much importance to it; and consequently, in the subsequent part of this work, though I shall occasionally refer to this cause of variation, I shall consider all the great variations which take place in the relative value of commodities to be produced by the greater or less quantity of labour which may be required from time to time to produce them.

It is hardly necessary to say that commodities which have the same quantity of labour bestowed on their production will differ in exchangeable value if they cannot be brought to market in the same time.

Suppose I employ twenty men at an expense of £1000 for a year in the production of a commodity, and at the end of the year I employ twenty men again for another year, at a further expense of £1000 in finishing or perfecting the same commodity, and that I bring it to market at the end of two years, if profits be 10 per cent., my commodity must sell for £2310; for I have employed £1000 capital for one year, and £2100 capital for one year more. Another man employs precisely the same quantity of labour, but he employs it all in the first year; he employs forty men at an expense of £2000, and at the end of the first year he sells it with 10 per cent. profit, or for £2200. Here, then, are two commodities having precisely the same quantity of labour bestowed on them, one of which sells for £2310—the other for £2200.

This case appears to differ from the last, but is, in fact, the same. In both cases the superior price of one commodity is owing to the greater length of time which must elapse before it can be brought to market. In the former case the machinery and cloth were more than double the value of the corn, although only double the quantity of labour was bestowed on them. In the second case, one commodity is more valuable than the other, although no more labour was employed on its production. The difference in value arises in both cases from the profits being accumulated as capital, and is only a just compensation for the time that the profits were withheld.

It appears, then, that the division of capital into different proportions of fixed and circulating capital, employed in different trades, introduces a considerable modification to the rule, which is of universal application when labour is almost exclusively employed in production; namely, that commodities never vary in value unless a greater or less quantity of labour be bestowed on their production, it being shown in this section that, without any variation in the quantity of labour, the rise of its value merely will occasion a fall in the exchangeable value of those goods in the production of which fixed capital is employed; the larger the amount of fixed capital, the greater will be the fall.

SECTION V

The principle that value does not vary with the rise or fall of wages, modified also by the unequal durability of capital, and by the unequal rapidity with which it is returned to its employer

In the last section we have supposed that, of two equal capitals, in two different occupations, the proportions of fixed and circulating capitals were unequal; now let us suppose them to be in the same proportion, but of unequal durability. In proportion as fixed capital is less durable it approaches to the nature of circulating capital. It will be consumed and its value reproduced in a shorter time, in order to preserve the capital of the manufacturer. We have just seen that in proportion as fixed capital preponderates in a manufacture, when wages rise the value of commodities produced in that manufacture is relatively lower than that of commodities produced in manufactures where circulating capital preponderates. In proportion to the less durability of fixed capital, and its approach to the nature of circulating capital, the same effect will be produced by the same cause.

If fixed capital be not of a durable nature it will require a great quantity of labour annually to keep it in its original state of efficiency; but the labour so bestowed may be considered as really expended on the commodity manufactured, which must bear a value in proportion to such labour. If I had a machine worth £20,000 which with very little labour was efficient to the production of commodities, and if the wear and tear of such machine were of trifling amount, and the general rate of profit 10 per cent., I should not require much more than £2000 to be added to the price of the goods, on account of the employment of my machine; but if the wear and tear of the machine were great, if the quantity of labour requisite to keep it in an efficient state were that of fifty men annually, I should require an additional price for my goods equal to that which would be obtained by any other manufacturer who employed fifty men in the production of other goods, and who used no machinery at all.

But a rise in the wages of labour would not equally affect commodities produced with machinery quickly consumed, and commodities produced with machinery slowly consumed. In the production of the one, a great deal of labour would be continually transferred to the commodity produced—in the other very little would be so transferred. Every rise of wages, therefore, or, which is the same thing, every fall of profits, would lower the relative value of those commodities which were produced with a capital of a durable nature, and would proportionally elevate those which were produced with capital more perishable. A fall of wages would have precisely the contrary effect.

I have already said that fixed capital is of various degrees of durability—suppose now a machine which could in any particular trade be employed to do the work of one hundred men for a year, and that it would last only for one year. Suppose, too, the machine to cost £5000, and the wages annually paid to one hundred men to be £5000, it is evident that it would be a matter of indifference to the manufacturer whether he bought the machine or employed the men. But suppose labour to rise, and consequently the wages of one hundred men for a year to amount to £5500, it is obvious that the manufacturer would now no longer hesitate, it would be for his interest to buy the machine and get his work done for £5000. But will not the machine rise in price, will not that also be worth £5500 in consequence of the rise of labour? It would rise in price if there were no stock employed on its construction, and no profits to be paid to the maker of it. If, for example, the machine were the produce of the labour of one hundred men, working one year upon it with wages of £50 each, and its price were consequently £5000; should those wages rise to £55, its price would be £5500, but this cannot be the case; less than one hundred men are employed or it could not be sold for £5000, for out of the £5000 must be paid the profits of stock which employed the men. Suppose then that only eighty-five men were employed at an expense of £50 each, or £4250 per annum, and that the £750 which the sale of the machine would produce over and above the wages advanced to the men constituted the profits of the engineer's stock. When wages rose 10 per cent., he would be obliged to employ an additional capital of £425, and would therefore employ £4675 instead of £4250, on which capital he would only get a profit of £325 if he continued to sell his machine for £5000; but this is precisely the case of all manufacturers and capitalists; the rise of wages affects them all. If therefore the maker of the machine should raise the price of it in consequence of a rise of wages, an unusual quantity of capital would be employed in the construction of such machines, till their price afforded only the common rate of profits.[1] We see then that machines would not rise in price in consequence of a rise of wages.

The manufacturer, however, who in a general rise of wages can have recourse to a machine which shall not increase the charge of production on his commodity, would enjoy peculiar advantages if he could continue to charge the same price for his goods; but he, as we have already seen, would be obliged to lower the price of his commodities, or capital would flow to his trade till his profits had sunk to the general level. Thus then is the public benefited by machinery: these mute agents are always the produce of much less labour than that which they displace, even when they are of the same money value. Through their influence an increase in the price of provisions which raises wages will affect fewer persons; it will reach, as in the above instance, eighty-five men instead of a hundred, and the saving which is the consequence shows itself in the reduced price of the commodity manufactured. Neither machines, nor the commodities made by them, rise in real value, but all commodities made by machines fall, and fall in proportion to their durability.

It will be seen then, that in the early stages of society, before much machinery or durable capital is used, the commodities produced by equal capitals will be nearly of equal value, and will rise or fall only relatively to each other on account of more or less labour being required for their production; but after the introduction of these expensive and durable instruments, the commodities produced by the employment of equal capitals will be of very unequal value, and although they will still be liable to rise or fall relatively to each other, as more or less labour becomes necessary to their production, they will be subject to another, though a minor variation, also from the rise

---

[1] We here see why it is that old countries are constantly impelled to employ machinery, and new countries to employ labour. With every difficulty of providing for the maintenance of men, labour necessarily rises, and with every rise in the price of labour, new temptations are offered to the use of machinery. This difficulty of providing for the maintenance of men is in constant operation in old countries; in new ones a very great increase in the population may take place without the least rise in the wages of labour. It may be as easy to provide for the seventh, eighth, and ninth million of men as for the second, third, and fourth.

or fall of wages and profits. Since goods which sell for £5000 may be the produce of a capital equal in amount to that from which are produced other goods which sell for £10,000, the profits on their manufacture will be the same; but those profits would be unequal if the prices of the goods did not vary with a rise or fall in the rate of profits.

It appears, too, that in proportion to the durability of capital employed in any kind of production the relative prices of those commodities on which such durable capital is employed will vary inversely as wages; they will fall as wages rise, and rise as wages fall; and, on the contrary, those which are produced chiefly by labour with less fixed capital, or with fixed capital of a less durable character than the medium in which price is estimated, will rise as wages rise, and fall as wages fall.

SECTION VI
On an invariable measure of value

WHEN commodities varied in relative value it would be desirable to have the means of ascertaining which of them fell and which rose in real value, and this could be effected only by comparing them one after another with some invariable standard measure of value, which should itself be subject to none of the fluctuations to which other commodities are exposed. Of such a measure it is impossible to be possessed, because there is no commodity which is not itself exposed to the same variations as the things the value of which is to be ascertained; that is, there is none which is not subject to require more or less labour for its production. But if this cause of variation in the value of a medium could be removed—if it were possible that in the production of our money, for instance, the same quantity of labour should at all times be required, still it would not be a perfect standard or invariable measure of value, because, as I have already endeavoured to explain, it would be subject to relative variations from a rise or fall of wages, on account of the different proportions of fixed capital which might be necessary to produce it, and to produce those other commodities whose alteration of value we wished to ascertain. It might be subject to variations, too, from the same cause, on account of the different degrees of durability of the fixed capital employed on it, and the commodities to be compared with it—or the time necessary to bring the one to market might be longer or shorter than the time necessary to bring the other commodities to market, the variations of which were to be determined; all which circumstances disqualify any commodity that can be thought of from being a perfectly accurate measure of value.

If, for example, we were to fix on gold as a standard, it is evident that it is but a commodity obtained under the same contingencies as every other commodity, and requiring labour and fixed capital to produce it. Like every other commodity, improvements in the saving of labour might be applied to its production, and consequently it might fall in relative value to other things merely on account of the greater facility of producing it.

If we suppose this cause of variation to be removed, and the same quantity of labour to be always required to obtain the same quantity of gold, still gold would not be a perfect measure of value, by which we could accurately ascertain the variations in all other things, because it would not be produced with precisely the same combinations of fixed and circulating capital as all other things; nor with fixed capital of the same durability; nor would it require precisely the same length of time before it could be brought to market. It would be a perfect measure of value for all things produced under the same circumstances precisely as itself, but for no others. If, for example, it were produced under the same circumstances as we have supposed necessary to produce cloth and cotton goods, it would be a perfect measure of value for those things, but not so for corn, for coals, and other commodities produced with either a less or a greater proportion of fixed capital, because, as we have shown, every alteration in the permanent rate of profits would have some effect on the relative value of all these goods, independently of any alteration in the quantity of labour employed on their production. If gold were produced under the same circumstances as corn, even if they never changed, it would not, for the same reasons, be at all times a perfect measure of the value of cloth and cotton goods. Neither gold, then, nor any other commodity, can ever be a perfect measure of value for all things; but I have already remarked that the effect on the relative prices of things, from a variation in profits, is comparatively slight; that by far the most important effects are produced by the varying quantities of labour required for production; and therefore, if we suppose this important cause of variation removed from the production of gold, we shall probably possess as near an approximation to a standard measure of value as can be theoretically conceived. May not gold be considered as a commodity produced with such proportions of the two kinds of capital as approach nearest to the average quantity employed in the production of most commodities? May not these proportions be so nearly equally distant from the two extremes, the one where little fixed capital is used, the other where little labour is employed, as to form a just mean between them?

If, then, I may suppose myself to be possessed of a standard so nearly approaching to an invariable one, the advantage is that I shall be enabled to speak of the variations of other things without embarrassing myself on every occasion with the consideration of the possible alteration in the value of the medium in which price and value are estimated.

To facilitate, then, the object of this inquiry, although I fully allow that money made of gold is subject to most of the variations of other things, I shall suppose it to be invariable, and therefore all alterations in price to be occasioned by some alteration in the value of the commodity of which I may be speaking.

Before I quit this subject, it may be proper to observe that Adam Smith, and all the writers who have followed him, have, without one exception that I know of, maintained that a rise in the price of labour would be uniformly followed by a rise in the price of all commodities. I hope I have succeeded in showing that there are no grounds for such an opinion, and that only those commodities would rise which had less fixed capital employed upon them than the medium in which price was estimated, and that all those which had more would positively fall in price when wages rose. On the contrary, if wages fell, those commodities only would fall which had a less proportion of fixed capital employed on them than the medium in which price was estimated; all those which had more would positively rise in price.

It is necessary for me also to remark that I have not said, because one commodity has so much labour bestowed upon it as will cost £1000, and another so much as will cost £2000, that therefore one would be of the value of £1000, and the other of the value of £2000; but I have said that their value will be to each other as two to one, and that in those proportions they will be exchanged. It is of no importance to the truth of this doctrine whether one of these commodities sells for £1100 and the other for £2200, or one for £1500 and the other for £3000; into that question I do not at present inquire; I affirm only that their relative values will be governed by the relative quantities of labour bestowed on their production.[1]

SECTION VII

Different effects from the alteration in the value of money, the medium in which PRICE is always expressed, or from the alteration in the value of the commodities which money purchases

ALTHOUGH I shall, as I have already explained, have occasion to consider money as invariable in value, for the purpose of more distinctly pointing out the causes of relative variations in the value of other things, it may be useful to notice the different effects which will follow from the prices of goods being altered by the causes to which I have already adverted, namely, the different quantities of labour required to produce them, and their being altered by a variation in the value of money itself.

Money being a variable commodity, the rise of money-wages will be frequently occasioned by a fall in the value of money. A rise of wages from this cause will, indeed, be invariably accompanied by a rise in the price of commodities; but in such cases it will be found that labour and all commodities have not varied in regard to each other, and that the variation has been confined to money.

Money, from its being a commodity obtained from a foreign country, from its being the general medium of exchange between all civilised countries, and from its being also distributed among those countries in proportions which are ever changing with every improvement in commerce and machinery, and with every increasing difficulty of obtaining food and necessaries for an increasing population, is subject to incessant variations. In stating the principles which regulate exchangeable value and price, we should carefully distinguish between those variations which belong to the commodity itself, and those which are occasioned by a variation in the medium in which value is estimated or price expressed.

[1] Mr. Malthus remarks on this doctrine, " We have the power indeed, arbitrarily, to call the labour which has been employed upon a commodity its real value, but in so doing we use words in a different sense from that in which they are customarily used; we confound at once the very important distinction between *cost* and *value* ; and render it almost impossible to explain with clearness the main stimulus to the production of wealth, which in fact depends upon this distinction."

Mr. Malthus appears to think that it is a part of my doctrine that the cost and value of a thing should be the same; it is, if he means by cost, " cost of production " including profits. In the above passage, this is what he does not mean, and therefore he has not clearly understood me.

A rise in wages, from an alteration in the value of money, produces a general effect on price, and for that reason it produces no real effect whatever on profits. On the contrary, a rise of wages, from the circumstance of the labourer being more liberally rewarded, or from a difficulty of procuring the necessaries on which wages are expended, does not, except in some instances, produce the effect of raising price, but has a great effect in lowering profits. In the one case, no greater proportion of the annual labour of the country is devoted to the support of the labourers; in the other case, a larger portion is so devoted.

It is according to the division of the whole produce of the land of any particular farm, between the three classes, of landlord, capitalist, and labourer, that we are to judge of the rise or fall of rent, profit, and wages, and not according to the value at which that produce may be estimated in a medium which is confessedly variable.

It is not by the absolute quantity of produce obtained by either class that we can correctly judge of the rate of profit, rent, and wages, but by the quantity of labour required to obtain that produce. By improvements in machinery and agriculture the whole produce may be doubled; but if wages, rent, and profit be also doubled, these three will bear the same proportions to one another as before, and neither could be said to have relatively varied. But if wages partook not of the whole of this increase; if they, instead of being doubled, were only increased one-half; if rent, instead of being doubled, were only increased three-fourths, and the remaining increase went to profit, it would, I apprehend, be correct for me to say that rent and wages had fallen while profits had risen; for if we had an invariable standard by which to measure the value of this produce we should find that a less value had fallen to the class of labourers and landlords, and a greater to the class of capitalists, than had been given before. We might find, for example, that though the absolute quantity of commodities had been doubled, they were the produce of precisely the former quantity of labour. Of every hundred hats, coats, and quarters of corn produced, if

The labourers had before	25
The landlords	25
And the capitalists	50
	100:

And if, after these commodities were double the quantity, of every 100

The labourers had only	22
The landlords	22
And the capitalists	56
	100:

In that case I should say that wages and rent had fallen and profits risen; though, in consequence of the abundance of commodities, the quantity paid to the labourer and landlord would have increased in the proportion of 25 to 44. Wages are to be estimated by their real value, viz., by the quantity of labour and capital employed in producing them, and not by their nominal value either in coats, hats, money, or corn. Under the circumstances I have just supposed, commodities would have fallen to half their former value, and if money had not varied, to half their former price also. If then in this medium, which had not varied in value, the wages of the labourer should be found to have fallen, it will not the less be a real fall because they might furnish him with a greater quantity of cheap commodities than his former wages.

The variation in the value of money, however great, makes no difference in the *rate* of profits; for suppose the goods of the manufacturer to rise from £1000 to £2000, or 100 per cent., if his capital, on which the variations of money have as much effect as on the value of produce, if his machinery, buildings, and stock in trade rise also a 100 per cent., his rate of profits will be the same, and he will have the same quantity, and no more, of the produce of the labour of the country at his command.

If, with a capital of a given value, he can, by economy in labour, double the quantity of produce, and it fall to half its former price, it will bear the same proportion to the capital that produced it which it did before, and consequently profits will still be at the same rate.

If, at the same time that he doubles the quantity of produce by the employment of the same capital, the value of money is by any accident lowered one half, the produce will sell for twice the money value that it did before; but the capital employed to produce it will also be of twice its former money value; and therefore in this case, too, the value of the produce will bear the same proportion to the value of the capital as it did before; and although the produce be doubled, rent, wages, and profits will only vary as the proportions vary, in which this double produce may be divided among the three classes that share it.

CHAPTER II

ON RENT

IT remains however to be considered whether the appropriation of land, and the consequent creation of rent, will occasion any variation in the relative value of commodities independently of the quantity of labour necessary to production. In order to understand this part of the subject we must inquire into the nature of rent, and the laws by which its rise or fall is regulated.

Rent is that portion of the produce of the earth which is paid to the landlord for the use of the original and indestructible powers of the soil. It is often, however, confounded with the interest and profit of capital, and, in popular language, the term is applied to whatever is annually paid by a farmer to his landlord. If, of two adjoining farms of the same extent, and of the same natural fertility, one had all the conveniences of farming buildings, and, besides, were properly drained and manured, and advantageously divided by hedges, fences, and walls, while the other had none of these advantages, more remuneration would naturally be paid for the use of one than for the use of the other; yet in both cases this remuneration would be called rent. But it is evident that a portion only of the money annually to be paid for the improved farm would be given for the original and indestructible powers of the soil; the other portion would be paid for the use of the capital which had been employed in ameliorating the quality of the land, and in erecting such buildings as were necessary to secure and preserve the produce. Adam Smith sometimes speaks of rent in the strict sense to which I am desirous of confining it, but more often in the popular sense in which the term is usually employed. He tells us that the demand for timber, and its consequent high price, in the more southern countries of Europe caused a rent to be paid for forests in Norway which could before afford no rent. Is it not, however, evident that the person who paid what he thus calls rent, paid it in consideration of the valuable commodity which was then standing on the land,

and that he actually repaid himself with a profit by the sale
of the timber? If, indeed, after the timber was removed, any
compensation were paid to the landlord for the use of the land,
for the purpose of growing timber or any other produce, with
a view to future demand, such compensation might justly be
called rent, because it would be paid for the productive powers
of the land; but in the case stated by Adam Smith, the compensation was paid for the liberty of removing and selling the
timber, and not for the liberty of growing it. He speaks also
of the rent of coal mines, and of stone quarries, to which the
same observation applies—that the compensation given for the
mine or quarry is paid for the value of the coal or stone which
can be removed from them, and has no connection with the
original and indestructible powers of the land. This is a distinction of great importance in an inquiry concerning rent and
profits; for it is found that the laws which regulate the progress
of rent are widely different from those which regulate the
progress of profits, and seldom operate in the same direction.
In all improved countries, that which is annually paid to the
landlord, partaking of both characters, rent and profit, is sometimes kept stationary by the effects of opposing causes; at
other times advances or recedes as one or the other of these
causes preponderates. In the future pages of this work, then,
whenever I speak of the rent of land, I wish to be understood
as speaking of that compensation which is paid to the owner of
land for the use of its original and indestructible powers.

On the first settling of a country in which there is an abundance of rich and fertile land, a very small proportion of which
is required to be cultivated for the support of the actual population, or indeed can be cultivated with the capital which the
population can command, there will be no rent; for no one
would pay for the use of land when there was an abundant
quantity not yet appropriated, and, therefore, at the disposal
of whosoever might choose to cultivate it.

On the common principles of supply and demand, no rent
could be paid for such land, for the reason stated why nothing
is given for the use of air and water, or for any other of the gifts
of nature which exist in boundless quantity. With a given
quantity of materials, and with the assistance of the pressure
of the atmosphere, and the elasticity of steam, engines may
perform work, and abridge human labour to a very great extent;
but no charge is made for the use of these natural aids, because
they are inexhaustible and at every man's disposal. In the

same manner, the brewer, the distiller, the dyer, make incessant use of the air and water for the production of their commodities; but as the supply is boundless, they bear no price.[1] If all land had the same properties, if it were unlimited in quantity, and uniform in quality, no charge could be made for its use, unless where it possessed peculiar advantages of situation. It is only, then, because land is not unlimited in quantity and uniform in quality, and because, in the progress of population, land of an inferior quality, or less advantageously situated, is called into cultivation, that rent is ever paid for the use of it. When, in the progress of society, land of the second degree of fertility is taken into cultivation, rent immediately commences on that of the first quality, and the amount of that rent will depend on the difference in the quality of these two portions of land.

When land of the third quality is taken into cultivation, rent immediately commences on the second, and it is regulated as before by the difference in their productive powers. At the same time, the rent of the first quality will rise, for that must always be above the rent of the second by the difference between the produce which they yield with a given quantity of capital and labour. With every step in the progress of population, which shall oblige a country to have recourse to land of a worse quality, to enable it to raise its supply of food, rent, on all the more fertile land, will rise.

Thus suppose land — No. 1, 2, 3 — to yield, with an equal employment of capital and labour, a net produce of 100, 90, and 80 quarters of corn. In a new country, where there is an abundance of fertile land compared with the population, and where therefore it is only necessary to cultivate No. 1, the whole net produce will belong to the cultivator, and will be the profits of the stock which he advances. As soon as population had so far increased as to make it necessary to cultivate No. 2, from which ninety quarters only can be obtained after supporting the labourers, rent would commence on No. 1; for either there must be two rates of profit on agricultural capital, or ten

[1] " The earth, as we have already seen, is not the only agent of nature which has a productive power; but it is the only one, or nearly so, that one set of men take to themselves to the exclusion of others; and of which, consequently, they can appropriate the benefits. The waters of rivers, and of the sea, by the power which they have of giving movement to our machines, carrying our boats, nourishing our fish, have also a productive power; the wind which turns our mills, and even the heat of the sun, work for us; but happily no one has yet been able to say, the ' wind and the sun are mine, and the service which they render must be paid for.' "—*Economie Politique*, par J. B. Say, vol. ii. p. 124.

quarters, or the value of ten quarters must be withdrawn from the produce of No. 1 for some other purpose. Whether the proprietor of the land, or any other person, cultivated No. 1, these ten quarters would equally constitute rent; for the cultivator of No. 2 would get the same result with his capital whether he cultivated No. 1, paying ten quarters for rent, or continued to cultivate No. 2, paying no rent. In the same manner it might be shown that when No. 3 is brought into cultivation, the rent of No. 2 must be ten quarters, or the value of ten quarters, whilst the rent of No. 1 would rise to twenty quarters; for the cultivator of No. 3 would have the same profits whether he paid twenty quarters for the rent of No. 1, ten quarters for the rent of No. 2, or cultivated No. 3 free of all rent.

It often, and, indeed, commonly happens, that before No. 2, 3, 4, or 5, or the inferior lands are cultivated, capital can be employed more productively on those lands which are already in cultivation. It may perhaps be found that by doubling the original capital employed on No. 1, though the produce will not be doubled, will not be increased by 100 quarters, it may be increased by eighty-five quarters, and that this quantity exceeds what could be obtained by employing the same capital on land No. 3.

In such case, capital will be preferably employed on the old land, and will equally create a rent; for rent is always the difference between the produce obtained by the employment of two equal quantities of capital and labour. If, with a capital of £1000 a tenant obtain 100 quarters of wheat from his land, and by the employment of a second capital of £1000 he obtain a further return of eighty-five, his landlord would have the power, at the expiration of his lease, of obliging him to pay fifteen quarters or an equivalent value for additional rent; for there cannot be two rates of profit. If he is satisfied with a diminution of fifteen quarters in the return for his second £1000, it is because no employment more profitable can be found for it. The common rate of profit would be in that proportion, and if the original tenant refused, some other person would be found willing to give all which exceeded that rate of profit to the owner of the land from which he derived it.

In this case, as well as in the other, the capital last employed pays no rent. For the greater productive powers of the first £1000, fifteen quarters, is paid for rent, for the employment of the second £1000 no rent whatever is paid. If a third £1000 be employed on the same land, with a return of seventy-five quarters, rent will then be paid for the second £1000, and will be equal to the difference between the produce of these two, or ten quarters; and at the same time the rent for the first £1000 will rise from fifteen to twenty-five quarters; while the last £1000 will pay no rent whatever.

If, then, good land existed in a quantity much more abundant than the production of food for an increasing population required, or if capital could be indefinitely employed without a diminished return on the old land, there could be no rise of rent; for rent invariably proceeds from the employment of an additional quantity of labour with a proportionally less return.

The most fertile and most favourably situated land will be first cultivated, and the exchangeable value of its produce will be adjusted in the same manner as the exchangeable value of all other commodities, by the total quantity of labour necessary in various forms, from first to last, to produce it and bring it to market. When land of an inferior quality is taken into cultivation, the exchangeable value of raw produce will rise, because more labour is required to produce it.

The exchangeable value of all commodities, whether they be manufactured, or the produce of the mines, or the produce of land, is always regulated, not by the less quantity of labour that will suffice for their production under circumstances highly favourable, and exclusively enjoyed by those who have peculiar facilities of production; but by the greater quantity of labour necessarily bestowed on their production by those who have no such facilities; by those who continue to produce them under the most unfavourable circumstances; meaning—by the most unfavourable circumstances, the most unfavourable under which the quantity of produce required renders it necessary to carry on the production.

Thus, in a charitable institution, where the poor are set to work with the funds of benefactors, the general prices of the commodities, which are the produce of such work, will not be governed by the peculiar facilities afforded to these workmen, but by the common, usual, and natural difficulties which every other manufacturer will have to encounter. The manufacturer enjoying none of these facilities might indeed be driven altogether from the market if the supply afforded by these favoured workmen were equal to all the wants of the community; but if he continued the trade, it would be only on condition that he should derive from it the usual and general rate of profits on stock; and that could only happen when his commodity sold for a price proportioned to the quantity of labour bestowed on its production.[1]

It is true, that on the best land, the same produce would still be obtained with the same labour as before, but its value would be enhanced in consequence of the diminished returns obtained by those who employed fresh labour and stock on the less fertile land. Notwithstanding, then, that the advantages of fertile over inferior lands are in no case lost, but only transferred from the cultivator, or consumer, to the landlord, yet, since more labour is required on the inferior lands, and since it is from such land only that we are enabled to furnish ourselves with the additional supply of raw produce, the comparative value of that produce will continue permanently above its former level, and make it exchange for more hats, cloth, shoes, etc., etc., in the production of which no such additional quantity of labour is required.

The reason, then, why raw produce rises in comparative value is because more labour is employed in the production of the last portion obtained, and not because a rent is paid to the landlord. The value of corn is regulated by the quantity of labour bestowed on its production on that quality of land, or with that portion of capital, which pays no rent. Corn is not high because a rent is paid, but a rent is paid because corn is high; and it has been justly observed that no reduction would take place in the price of corn although landlords should forego the whole of their rent. Such a measure would only enable some farmers to live like gentlemen, but would not diminish

---

[1] Has not M. Say forgotten, in the following passage, that it is the cost of production which ultimately regulates price? " The produce of labour employed on the land has this peculiar property, that it does not become more dear by becoming more scarce, because population always diminishes at the same time that food diminishes, and consequently the quantity of these products *demanded* diminishes at the same time as the quantity supplied. Besides, it is not observed that corn is more dear in those places where there is plenty of uncultivated land, than in completely cultivated countries. England and France were much more imperfectly cultivated in the middle ages than they are now; they produced much less raw produce: nevertheless, from all that we can judge by a comparison with the value of other things, corn was not sold at a dearer price. If the produce was less, so was the population; the weakness of the demand compensated the feebleness of the supply " (vol. ii. 338). M. Say being impressed with the opinion that the price of commodities is regulated by the price of labour, and justly supposing that charitable institutions of all sorts tend to increase the population beyond what it otherwise would be, and therefore to lower wages, says, " I suspect that the cheapness of the goods which come from England is partly caused by the numerous charitable institutions which exist in that country " (vol. ii. 277). This is a consistent opinion in one who maintains that wages regulate price.

the quantity of labour necessary to raise raw produce on the least productive land in cultivation.

Nothing is more common than to hear of the advantages which the land possesses over every other source of useful produce, on account of the surplus which it yields in the form of rent. Yet when land is most abundant, when most productive, and most fertile, it yields no rent; and it is only when its powers decay, and less is yielded in return for labour, that a share of the original produce of the more fertile portions is set apart for rent. It is singular that this quality in the land, which should have been noticed as an imperfection compared with the natural agents by which manufacturers are assisted, should have been pointed out as constituting its peculiar pre-eminence. If air, water, the elasticity of steam, and the pressure of the atmosphere were of various qualities; if they could be appropriated, and each quality existed only in moderate abundance, they, as well as the land, would afford a rent, as the successive qualities were brought into use. With every worse quality employed, the value of the commodities in the manufacture of which they were used would rise, because equal quantities of labour would be less productive. Man would do more by the sweat of his brow and nature perform less; and the land would be no longer pre-eminent for its limited powers.

If the surplus produce which land affords in the form of rent be an advantage, it is desirable that, every year, the machinery newly constructed should be less efficient than the old, as that would undoubtedly give a greater exchangeable value to the goods manufactured, not only by that machinery but by all the other machinery in the kingdom; and a rent would be paid to all those who possessed the most productive machinery.[1]

---

[1] " In agriculture, too," says Adam Smith, " nature labours along with man; and though her labour costs no expense, its produce has its value, as well as that of the most expensive workman." The labour of nature is paid, not because she does much, but because she does little. In proportion as she becomes niggardly in her gifts she exacts a greater price for her work. Where she is munificently beneficent she always works gratis. " The labouring cattle employed in agriculture not only occasion, like the workmen in manufactures, the reproduction of a value equal to their own consumption, or to the capital which employs them, together with its owner's profits, but of a much greater value. Over and above the capital of the farmer and all its profits, they regularly occasion the reproduction of the rent of the landlord. This rent may be considered as the produce of those powers of nature, the use of which the landlord lends to the farmer. It is greater or smaller according to the supposed extent of those powers, or, in other words, according to the supposed natural or improved fertility of the land. It is the work of nature which remains, after deducting or compensating everything which can be regarded as the work of man. It

The rise of rent is always the effect of the increasing wealth of the country, and of the difficulty of providing food for its augmented population. It is a symptom, but it is never a cause of wealth; for wealth often increases most rapidly while rent is either stationary, or even falling. Rent increases most rapidly as the disposable land decreases in its productive powers. Wealth increases most rapidly in those countries where the disposable land is most fertile, where importation is least restricted, and where, through agricultural improvements, productions can be multiplied without any increase in the proportional quantity of labour, and where consequently the progress of rent is slow.

If the high price of corn were the effect, and not the cause of rent, price would be proportionally influenced as rents were high or low, and rent would be a component part of price. But that corn which is produced by the greatest quantity of labour

is seldom less than a fourth, and frequently more than a third of the whole produce. No equal quantity of productive labour employed in manufactures can ever occasion so great a reproduction. *In them nature does nothing, man does all ;* and the reproduction must always be in proportion to the strength of the agents that occasion it. The capital employed in agriculture, therefore, not only puts into motion a greater quantity of productive labour than any equal capital employed in manufactures, but in proportion, too, to the quantity of the productive labour which it employs it adds a much greater value to the annual produce of the land and labour of the country, to the *real* wealth and revenue of its inhabitants. Of all the ways in which a capital can be employed, it is by far the most advantageous to the society."—Book II. chap. v. p. 15.

Does nature nothing for man in manufactures? Are the powers of wind and water, which move our machinery and assist navigation, nothing? The pressure of the atmosphere and the elasticity of steam, which enable us to work the most stupendous engines—are they not the gifts of nature? To say nothing of the effects of the matter of heat in softening and melting metals, of the decomposition of the atmosphere in the process of dyeing and fermentation. There is not a manufacture which can be mentioned in which nature does not give her assistance to man, and give it, too, generously and gratuitously.

In remarking on the passage which I have copied from Adam Smith, Mr. Buchanan observes, " I have endeavoured to show, in the observations on productive and unproductive labour, contained in the fourth volume, that agriculture adds no more to the national stock than any other sort of industry. In dwelling on the reproduction of rent as so great an advantage to society, Dr. Smith does not reflect that rent is the effect of high price, and that what the landlord gains in this way he gains at the expense of the community at large. There is no absolute gain to the society by the reproduction of rent; it is only one class profiting at the expense of another class. The notion of agriculture yielding a produce, and a rent in consequence, because nature concurs with human industry in the process of cultivation, is a mere fancy. It is not from the produce, but from the price at which the produce is sold, that the rent is derived; and this price is got not because nature assists in the production, but because it is the price which suits the consumption to the supply.

is the regulator of the price of corn; and rent does not and cannot enter in the least degree as a component part of its price.[1] Adam Smith, therefore, cannot be correct in supposing that the original rule which regulated the exchangeable value of commodities, namely, the comparative quantity of labour by which they were produced, can be at all altered by the appropriation of land and the payment of rent. Raw material enters into the composition of most commodities, but the value of that raw material, as well as corn, is regulated by the productiveness of the portion of capital last employed on the land and paying no rent; and therefore rent is not a component part of the price of commodities.

We have been hitherto considering the effects of the natural progress of wealth and population on rent in a country in which the land is of variously productive powers, and we have seen that with every portion of additional capital which it becomes necessary to employ on the land with a less productive return rent would rise. It follows from the same principles that any circumstances in the society which should make it unnecessary to employ the same amount of capital on the land, and which should therefore make the portion last employed more productive, would lower rent. Any great reduction in the capital of a country which should materially diminish the funds destined for the maintenance of labour, would naturally have this effect. Population regulates itself by the funds which are to employ it, and therefore always increases or diminishes with the increase or diminution of capital. Every reduction of capital is therefore necessarily followed by a less effective demand for corn, by a fall of price, and by diminished cultivation. In the reverse order to that in which the accumulation of capital raises rent will the diminution of it lower rent. Land of a less unproductive quality will be in succession relinquished, the exchangeable value of produce will fall, and land of a superior quality will be the land last cultivated, and that which will then pay no rent.

The same effects may, however, be produced when the wealth and population of a country are increased, if that increase is accompanied by such marked improvements in agriculture as shall have the same effect of diminishing the necessity of cultivating the poorer lands, or of expending the same amount of capital on the cultivation of the more fertile portions.

[1] The clearly understanding this principle is, I am persuaded, of the utmost importance to the science of political economy.

If a million of quarters of corn be necessary for the support of a given population, and it be raised on land of the qualities of No. 1, 2, 3; and if an improvement be afterwards discovered by which it can be raised on No. 1 and 2, without employing No. 3, it is evident that the immediate effect must be a fall of rent; for No. 2, instead of No. 3, will then be cultivated without paying any rent; and the rent of No. 1, instead of being the difference between the produce of No. 3 and No. 1, will be the difference only between No. 2 and 1. With the same population, and no more, there can be no demand for any additional quantity of corn; the capital and labour employed on No. 3 will be devoted to the production of other commodities desirable to the community, and can have no effect in raising rent, unless the raw material from which they are made cannot be obtained without employing capital less advantageously on the land, in which case No. 3 must again be cultivated.

It is undoubtedly true that the fall in the relative price of raw produce, in consequence of the improvement in agriculture, or rather in consequence of less labour being bestowed on its production, would naturally lead to increased accumulation; for the profits of stock would be greatly augmented. This accumulation would lead to an increased demand for labour, to higher wages, to an increased population, to a further demand for raw produce, and to an increased cultivation. It is only, however, after the increase in the population that rent would be as high as before; that is to say, after No. 3 was taken into cultivation. A considerable period would have elapsed, attended with a positive diminution of rent.

But improvements in agriculture are of two kinds: those which increase the productive powers of the land and those which enable us, by improving our machinery, to obtain its produce with less labour. They both lead to a fall in the price of raw produce; they both affect rent, but they do not affect it equally. If they did not occasion a fall in the price of raw produce they would not be improvements; for it is the essential quality of an improvement to diminish the quantity of labour before required to produce a commodity; and this diminution cannot take place without a fall of its price or relative value.

The improvements which increased the productive powers of the land are such as the more skilful rotation of crops or the better choice of manure. These improvements absolutely enable us to obtain the same produce from a smaller quantity of land. If, by the introduction of a course of turnips, I can

feed my sheep besides raising my corn, the land on which the
sheep were before fed becomes unnecessary, and the same
quantity of raw produce is raised by the employment of a less
quantity of land. If I discover a manure which will enable me
to make a piece of land produce 20 per cent. more corn, I may
withdraw at least a portion of my capital from the most unproductive part of my farm. But, as I before observed, it is not
necessary that land should be thrown out of cultivation in order
to reduce rent: to produce this effect, it is sufficient that successive portions of capital are employed on the same land with
different results, and that the portion which gives the least
result should be withdrawn. If, by the introduction of the
turnip husbandry, or by the use of a more invigorating manure,
I can obtain the same produce with less capital, and without
disturbing the difference between the productive powers of the
successive portions of capital, I shall lower rent; for a different
and more productive portion will be that which will form the
standard from which every other will be reckoned. If, for
example, the successive portions of capital yielded 100, 90, 80,
70; whilst I employed these four portions, my rent would be
60, or the difference between

$$
\left.\begin{array}{l}
70 \text{ and } 100 = 30 \\
70 \text{ and } 90 = 20 \\
70 \text{ and } 80 = 10 \\
\overline{60}
\end{array}\right\} \text{ whilst the produce would be } 340 \left\{\begin{array}{l}
100 \\
90 \\
80 \\
70 \\
\overline{340}
\end{array}\right.
$$

and while I employed these portions, the rent would remain
the same, although the produce of each should have an equal
augmentation. If, instead of 100, 90, 80, 70, the produce
should be increased to 125, 115, 105, 95, the rent would still
be 60, or the difference between

$$
\left.\begin{array}{l}
95 \text{ and } 125 = 30 \\
95 \text{ and } 115 = 20 \\
95 \text{ and } 105 = 10 \\
\overline{60}
\end{array}\right\} \begin{array}{l} \text{whilst the produce would be} \\ \text{increased to } 440 \end{array} \left\{\begin{array}{l}
125 \\
115 \\
105 \\
95 \\
\overline{440}
\end{array}\right.
$$

But with such an increase of produce, without an increase of
demand,[1] there could be no motive for employing so much

---

[1] I hope I am not understood as undervaluing the importance of all sorts
of improvements in agriculture to landlords—their immediate effect is to
lower rent; but as they give a great stimulus to population, and at the
same time enable us to cultivate poorer lands with less labour, they are
ultimately of immense advantage to landlords. A period, however, must
elapse during which they are positively injurious to him ⸱

capital on the land; one portion would be withdrawn, and consequently the last portion of capital would yield 105 instead of 95, and rent would fall to 30, or the difference between

$$
\left.\begin{array}{l}
105 \text{ and } 125 = 20 \\
105 \text{ and } 115 = 10 \\
\underline{} \\
30
\end{array}\right\}
\begin{array}{l}
\text{whilst the produce will be still} \\
\text{adequate to the wants of the popula-} \\
\text{tion, for it would be 345 quarters,} \\
\text{or}
\end{array}
\left\{
\begin{array}{l}
125 \\
115 \\
105 \\
\underline{} \\
345
\end{array}\right.
$$

the demand being only for 340 quarters.—But there are improvements which may lower the relative value of produce without lowering the corn rent, though they will lower the money rent of land. Such improvements do not increase the productive powers of the land, but they enable us to obtain its produce with less labour. They are rather directed to the formation of the capital applied to the land than to the cultivation of the land itself. Improvements in agricultural implements, such as the plough and the thrashing machine, economy in the use of horses employed in husbandry, and a better knowledge of the veterinary art, are of this nature. Less capital, which is the same thing as less labour, will be employed on the land; but to obtain the same produce, less land cannot be cultivated. Whether improvements of this kind, however, affect corn rent, must depend on the question whether the difference between the produce obtained by the employment of different portions of capital be increased, stationary, or diminished. If four portions of capital, 50, 60, 70, 80, be employed on the land, giving each the same results, and any improvement in the formation of such capital should enable me to withdraw 5 from each, so that they should be 45, 55, 65, and 75, no alteration would take place in the corn rent; but if the improvements were such as to enable me to make the whole saving on that portion of capital which is least productively employed, corn rent would immediately fall, because the difference between the capital most productive and the capital least productive would be diminished; and it is this difference which constitutes rent.

Without multiplying instances, I hope enough has been said to show that whatever diminishes the inequality in the produce obtained from successive portions of capital employed on the same or on new land tends to lower rent; and that whatever increases that inequality, necessarily produces an opposite effect, and tends to raise it.

In speaking of the rent of the landlord, we have rather considered it as the proportion of the produce, obtained with a given capital on any given farm, without any reference to its exchangeable value; but since the same cause, the difficulty of production, raises the exchangeable value of raw produce, and raises also the proportion of raw produce paid to the landlord for rent, it is obvious that the landlord is doubly benefited by difficulty of production. First, he obtains a greater share, and, secondly, the commodity in which he is paid is of greater value.[1]

[1] To make this obvious, and to show the degrees in which corn and money rent will vary, let us suppose that the labour of ten men will, on land of a certain quality, obtain 180 quarters of wheat, and its value to be £4 per quarter, or £720; and that the labour of ten additional men will, on the same or any other land, produce only 170 quarters in addition; wheat would rise from £4 to £4 4s. 8d. for 170: 180: : £4: £4 4s. 8d.; or, as in the production of 170 quarters, the labour of 10 men is necessary in one case, and only of 9.44 in the other, the rise would be as 9.44 to 10, or as £4 to £4 4s. 8d. If 10 men be further employed, and the return be

160	the price will rise to	£4	10	0
150	,,	,,	4	16	0
140	,,	,,	5	2	10

Now, if no rent was paid for the land which yielded 180 quarters, when corn was at £4 per quarter, the value of 10 quarters would be paid as rent when only 170 could be procured, which at £4 4s. 8d. would £42 7s. 6d.

20 quarters	when 160 were produced, which at	£4	10	0	would be	£90	0	0
30 quarters	,, 150	,,	4	16	0	,,	144	0	0
40 quarters	,, 140	,,	5	2	10	,,	205	13	4

Corn rent would increase in the proportion of $\begin{cases} 100 \\ 200 \\ 300 \\ 400 \end{cases}$ and money rent in the proportion of $\begin{cases} 100 \\ 212 \\ 340 \\ 485 \end{cases}$

CHAPTER III

ON THE RENT OF MINES

THE metals, like other things, are obtained by labour. Nature, indeed, produces them; but it is the labour of man which extracts them from the bowels of the earth and prepares them for our service.

Mines, as well as land, generally pay a rent to their owner; and this rent, as well as the rent of land, is the effect and never the cause of the high value of their produce.

If there were abundance of equally fertile mines, which any one might appropriate, they could yield no rent; the value of their produce would depend on the quantity of labour necessary to extract the metal from the mine and bring it to market.

But there are mines of various qualities affording very different results with equal quantities of labour. The metal produced from the poorest mine that is worked must at least have an exchangeable value, not only sufficient to procure all the clothes, food, and other necessaries consumed by those employed in working it, and bringing the produce to market, but also to afford the common and ordinary profits to him who advances the stock necessary to carry on the undertaking. The return for capital from the poorest mine paying no rent would regulate the rent of all the other more productive mines. This mine is supposed to yield the usual profits of stock. All that the other mines produce more than this will necessarily be paid to the owners for rent. Since this principle is precisely the same as that which we have already laid down respecting land, it will not be necessary further to enlarge on it.

It will be sufficient to remark that the same general rule which regulates the value of raw produce and manufactured commodities is applicable also to the metals; their value depending not on the rate of profits, nor on the rate of wages, nor on the rent paid for mines, but on the total quantity of labour necessary to obtain the metal and to bring it to market.

Like every other commodity, the value of the metals is subject to variation. Improvements may be made in the implements and machinery used in mining, which may considerably abridge labour; new and more productive mines may be discovered, in which, with the same labour, more metal may be obtained;

or the facilities of bringing it to market may be increased. In either of these cases the metals would fall in value, and would therefore exchange for a less quantity of other things. On the other hand, from the increasing difficulty of obtaining the metal, occasioned by the greater depth at which the mine must be worked, and the accumulation of water, or any other contingency, its value compared with that of other things might be considerably increased.

It has therefore been justly observed that however honestly the coin of a country may conform to its standard, money made of gold and silver is still liable to fluctuations in value, not only to accidental and temporary, but to permanent and natural variations, in the same manner as other commodities.

By the discovery of America, and the rich mines in which it abounds, a very great effect was produced on the natural price of the precious metals. This effect is by many supposed not yet to have terminated. It is probable, however, that all the effects on the value of the metals resulting from the discovery of America have long ceased; and if any fall has of late years taken place in their value, it is to be attributed to improvements in the mode of working the mines.

From whatever cause it may have proceeded, the effect has been so slow and gradual that little practical inconvenience has been felt from gold and silver being the general medium in which the value of all other things is estimated. Though undoubtedly a variable measure of value, there is probably no commodity subject to fewer variations. This and the other advantages which these metals possess, such as their hardness, their malleability, their divisibility, and many more, have justly secured the preference everywhere given to them as a standard for the money of civilised countries.

If equal quantities of labour, with equal quantities of fixed capital, could at all times obtain from that mine which paid no rent equal quantities of gold, gold would be as nearly an invariable measure of value as we could in the nature of things possess. The quantity indeed would enlarge with the demand, but its value would be invariable, and it would be eminently well calculated to measure the varying value of all other things. I have already in a former part of this work considered gold as endowed with this uniformity, and in the following chapter I shall continue the supposition. In speaking therefore of varying price, the variation will be always considered as being in the commodity, and never in the medium in which it is estimated.

CHAPTER IV

ON NATURAL AND MARKET PRICE

In making labour the foundation of the value of commodities, and the comparative quantity of labour which is necessary to their production, the rule which determines the respective quantities of goods which shall be given in exchange for each other, we must not be supposed to deny the accidental and temporary deviations of the actual or market price of commodities from this, their primary and natural price.

In the ordinary course of events, there is no commodity which continues for any length of time to be supplied precisely in that degree of abundance which the wants and wishes of mankind require, and therefore there is none which is not subject to accidental and temporary variations of price.

It is only in consequence of such variations that capital is apportioned precisely, in the requisite abundance and no more, to the production of the different commodities which happen to be in demand. With the rise or fall of price, profits are elevated above, or depressed below, their general level; and capital is either encouraged to enter into, or is warned to depart from, the particular employment in which the variation has taken place.

Whilst every man is free to employ his capital where he pleases, he will naturally seek for it that employment which is most advantageous; he will naturally be dissatisfied with a profit of 10 per cent., if by removing his capital he can obtain a profit of 15 per cent. This restless desire on the part of all the employers of stock to quit a less profitable for a more advantageous business has a strong tendency to equalise the rate of profits of all, or to fix them in such proportions as may, in the estimation of the parties, compensate for any advantage which one may have, or may appear to have, over the other. It is perhaps very difficult to trace the steps by which this change is effected: it is probably effected by a manufacturer not absolutely changing his employment, but only lessening the quantity of capital he has in that employment. In all rich countries there is a number of men forming what is called the moneyed class; these men are engaged in no trade, but live on the interest of their money, which is employed in discounting bills, or in loans to the more industrious part of the community. The bankers too employ a large capital on the same objects. The capital so employed forms a circulating capital of a large amount, and is employed, in larger or smaller proportions, by all the different trades of a country. There is perhaps no manufacturer, however rich, who limits his business to the extent that his own funds alone will allow: he has always some portion of this floating capital, increasing or diminishing according to the activity of the demand for his commodities. When the demand for silks increases, and that for cloth diminishes, the clothier does not remove with his capital to the silk trade, but he dismisses some of his workmen, he discontinues his demand for the loan from bankers and moneyed men; while the case of the silk manufacturer is the reverse: he wishes to employ more workmen, and thus his motive for borrowing is increased; he borrows more, and thus capital is transferred from one employment to another without the necessity of a manufacturer discontinuing his usual occupation. When we look to the markets of a large town, and observe how regularly they are supplied both with home and foreign commodities, in the quantity in which they are required, under all the circumstances of varying demand, arising from the caprice of taste, or a change in the amount of population, without often producing either the effects of a glut from a too abundant supply, or an enormously high price from the supply being unequal to the demand, we must confess that the principle which apportions capital to each trade in the precise amount that it is required is more active than is generally supposed.

A capitalist, in seeking profitable employment for his funds, will naturally take into consideration all the advantages which one occupation possesses over another. He may therefore be willing to forego a part of his money profit in consideration of the security, cleanliness, ease, or any other real or fancied advantage which one employment may possess over another.

If from a consideration of these circumstances, the profits of stock should be so adjusted that in one trade they were 20, in another 25, and in another 30 per cent., they would probably continue permanently with that relative difference, and with that difference only; for if any cause should elevate the profits of one of these trades 10 per cent., either these profits would be temporary, and would soon again fall back to their usual station, or the profits of the others would be elevated in the same proportion.

The present time appears to be one of the exceptions to the justness of this remark. The termination of the war has so deranged the division which before existed of employments in Europe, that every capitalist has not yet found his place in the new division which has now become necessary.

Let us suppose that all commodities are at their natural price, and consequently that the profits of capital in all employments are exactly at the same rate, or differ only so much as, in the estimation of the parties, is equivalent to any real or fancied advantage which they possess or forego. Suppose now that a change of fashion should increase the demand for silks and lessen that for woollens; their natural price, the quantity of labour necessary to their production, would continue unaltered, but the market price of silks would rise and that of woollens would fall; and consequently the profits of the silk manufacturer would be above, whilst those of the woollen manufacturer would be below, the general and adjusted rate of profits. Not only the profits, but the wages of the workmen, would be affected in these employments. This increased demand for silks would, however, soon be supplied by the transference of capital and labour from the woollen to the silk manufacture; when the market prices of silks and woollens would again approach their natural prices, and then the usual profits would be obtained by the respective manufacturers of those commodities.

It is then the desire, which every capitalist has, of diverting his funds from a less to a more profitable employment that prevents the market price of commodities from continuing for any length of time either much above or much below their natural price. It is this competition which so adjusts the changeable value of commodities that, after paying the wages for the labour necessary to their production, and all other expenses required to put the capital employed in its original state of efficiency, the remaining value or overplus will in each trade be in proportion to the value of the capital employed.

In the seventh chapter of the *Wealth of Nations*, all that concerns this question is most ably treated. Having fully acknowledged the temporary effects which, in particular employments of capital, may be produced on the prices of commodities, as well as on the wages of labour, and the profits of stock, by accidental causes, without influencing the general price of com-

modities, wages, or profits, since these effects are equally opera-
tive in all stages of society, we will leave them entirely out of
our consideration whilst we are treating of the laws which
regulate natural prices, natural wages, and natural profits,
effects totally independent of these accidental causes. In
speaking, then, of the exchangeable value of commodities, or
the power of purchasing possessed by any one commodity, I
mean always that power which it would possess if not disturbed
by any temporary or accidental cause, and which is its natural
price.

CHAPTER V

ON WAGES

LABOUR, like all other things which are purchased and sold, and which may be increased or diminished in quantity, has its natural and its market price. The natural price of labour is that price which is necessary to enable the labourers, one with another, to subsist and to perpetuate their race, without either increase or diminution.

The power of the labourer to support himself, and the family which may be necessary to keep up the number of labourers, does not depend on the quantity of money which he may receive for wages, but on the quantity of food, necessaries, and conveniences become essential to him from habit which that money will purchase. The natural price of labour, therefore, depends on the price of the food, necessaries, and conveniences required for the support of the labourer and his family. With a rise in the price of food and necessaries, the natural price of labour will rise; with the fall in their price, the natural price of labour will fall.

With the progress of society the natural price of labour has always a tendency to rise, because one of the principal commodities by which its natural price is regulated has a tendency to become dearer from the greater difficulty of producing it. As, however, the improvements in agriculture, the discovery of new markets, whence provisions may be imported, may for a time counteract the tendency to a rise in the price of necessaries, and may even occasion their natural price to fall, so will the same causes produce the correspondent effects on the natural price of labour.

The natural price of all commodities, excepting raw produce and labour, has a tendency to fall in the progress of wealth and population; for though, on one hand, they are enhanced in real value, from the rise in the natural price of the raw material of which they are made, this is more than counterbalanced by the improvements in machinery, by the better division and distribution of labour, and by the increasing skill, both in science and art, of the producers.

52

The market price of labour is the price which is really paid for it, from the natural operation of the proportion of the supply to the demand; labour is dear when it is scarce and cheap when it is plentiful. However much the market price of labour may deviate from its natural price, it has, like commodities, a tendency to conform to it.

It is when the market price of labour exceeds its natural price that the condition of the labourer is flourishing and happy, that he has it in his power to command a greater proportion of the necessaries and enjoyments of life, and therefore to rear a healthy and numerous family. When, however, by the encouragement which high wages give to the increase of population, the number of labourers is increased, wages again fall to their natural price, and indeed from a reaction sometimes fall below it.

When the market price of labour is below its natural price, the condition of the labourers is most wretched: then poverty deprives them of those comforts which custom renders absolute necessaries. It is only after their privations have reduced their number, or the demand for labour has increased, that the market price of labour will rise to its natural price, and that the labourer will have the moderate comforts which the natural rate of wages will afford.

Notwithstanding the tendency of wages to conform to their natural rate, their market rate may, in an improving society, for an indefinite period, be constantly above it; for no sooner may the impulse which an increased capital gives to a new demand for labour be obeyed, than another increase of capital may produce the same effect; and thus, if the increase of capital be gradual and constant, the demand for labour may give a continued stimulus to an increase of people.

Capital is that part of the wealth of a country which is employed in production, and consists of food, clothing, tools, raw materials, machinery, etc., necessary to give effect to labour.

Capital may increase in quantity at the same time that its value rises. An addition may be made to the food and clothing of a country at the same time that more labour may be required to produce the additional quantity than before; in that case not only the quantity but the value of capital will rise.

Or capital may increase without its value increasing, and even while its value is actually diminishing; not only may an addition be made to the food and clothing of a country, but the addition may be made by the aid of machinery, without any increase, and even with an absolute diminution in the proportional quantity of labour required to produce them. The quantity of capital may increase, while neither the whole together, nor any part of it singly, will have a greater value than before, but may actually have a less.

In the first case, the natural price of labour, which always depends on the price of food, clothing, and other necessaries, will rise; in the second, it will remain stationary or fall; but in both cases the market rate of wages will rise, for in proportion to the increase of capital will be the increase in the demand for labour; in proportion to the work to be done will be the demand for those who are to do it.

In both cases, too, the market price of labour will rise above its natural price; and in both cases it will have a tendency to conform to its natural price, but in the first case this agreement will be most speedily effected. The situation of the labourer will be improved, but not much improved; for the increased price of food and necessaries will absorb a large portion of his increased wages; consequently a small supply of labour, or a trifling increase in the population, will soon reduce the market price to the then increased natural price of labour.

In the second case, the condition of the labourer will be very greatly improved; he will receive increased money wages without having to pay any increased price, and perhaps even a diminished price for the commodities which he and his family consume; and it will not be till after a great addition has been made to the population that the market price of labour will again sink to its then low and reduced natural price.

Thus, then, with every improvement of society, with every increase in its capital, the market wages of labour will rise; but the permanence of their rise will depend on the question whether the natural price of labour has also risen; and this again will depend on the rise in the natural price of those necessaries on which the wages of labour are expended.

It is not to be understood that the natural price of labour, estimated even in food and necessaries, is absolutely fixed and constant. It varies at different times in the same country, and very materially differs in different countries.[1] It essentially

[1] " The shelter and the clothing which are indispensable in one country may be no way necessary in another; and a labourer in Hindostan may continue to work with perfect vigour, though receiving, as his natural wages, only such a supply of covering as would be insufficient to preserve a labourer in Russia from perishing. Even in countries situated in the same climate, different habits of living will often occasion variations in the

depends on the habits and customs of the people. An English labourer would consider his wages under their natural rate, and too scanty to support a family, if they enabled him to purchase no other food than potatoes, and to live in no better habitation than a mud cabin; yet these moderate demands of nature are often deemed sufficient in countries where " man's life is cheap " and his wants easily satisfied. Many of the conveniences now enjoyed in an English cottage would have been thought luxuries at an earlier period of our history.

From manufactured commodities always falling and raw produce always rising, with the progress of society, such a disproportion in their relative value is at length created, that in rich countries a labourer, by the sacrifice of a very small quantity only of his food, is able to provide liberally for all his other wants.

Independently of the variations in the value of money, which necessarily affect money wages, but which we have here supposed to have no operation, as we have considered money to be uniformly of the same value, it appears then that wages are subject to a rise or fall from two causes:—

First, the supply and demand of labourers.

Secondly, the price of the commodities on which the wages of labour are expended.

In different stages of society, the accumulation of capital, or of the means of employing labour, is more or less rapid, and must in all cases depend on the productive powers of labour. The productive powers of labour are generally greatest when there is an abundance of fertile land: at such periods accumulation is often so rapid that labourers cannot be supplied with the same rapidity as capital.

It has been calculated that under favourable circumstances population may be doubled in twenty-five years; but under the same favourable circumstances the whole capital of a country might possibly be doubled in a shorter period. In that case, wages during the whole period would have a tendency to rise, because the demand for labour would increase still faster than the supply.

In new settlements, where the arts and knowledge of countries far advanced in refinement are introduced, it is probable that capital has a tendency to increase faster than mankind; and

natural price of labour as considerable as those which are produced by natural causes."—P. 68. *An Essay on the External Corn Trade*, by R. Torrens, Esq.

The whole of this subject is most ably illustrated by Colonel Torrens.

if the deficiency of labourers were not supplied by more populous countries, this tendency would very much raise the price of labour. In proportion as these countries become populous, and land of a worse quality is taken into cultivation, the tendency to an increase of capital diminishes; for the surplus produce remaining, after satisfying the wants of the existing population, must necessarily be in proportion to the facility of production, viz. to the smaller number of persons employed in production. Although, then, it is probable that, under the most favourable circumstances, the power of production is still greater than that of population, it will not long continue so; for the land being limited in quantity, and differing in quality, with every increased portion of capital employed on it there will be a decreased rate of production, whilst the power of population continues always the same.

In those countries where there is abundance of fertile land, but where, from the ignorance, indolence, and barbarism of the inhabitants, they are exposed to all the evils of want and famine, and where it has been said that population presses against the means of subsistence, a very different remedy should be applied from that which is necessary in long settled countries, where, from the diminishing rate of the supply of raw produce, all the evils of a crowded population are experienced. In the one case, the evil proceeds from bad government, from the insecurity of property, and from a want of education in all ranks of the people. To be made happier they require only to be better governed and instructed, as the augmentation of capital, beyond the augmentation of people, would be the inevitable result. No increase in the population can be too great, as the powers of production are still greater. In the other case, the population increases faster than the funds required for its support. Every exertion of industry, unless accompanied by a diminished rate of increase in the population, will add to the evil, for production cannot keep pace with it.

With a population pressing against the means of subsistence, the only remedies are either a reduction of people or a more rapid accumulation a capital. In rich countries, where all the fertile land is already cultivated, the latter remedy is neither very practicable nor very desirable, because its effort would be, if pushed very far, to render all classes equally poor. But in poor countries, where there are abundant means of production in store, from fertile land not yet brought into cultivation, it is the only safe and efficacious means of removing the evil, particularly as its effect would be to elevate all classes of the people.

The friends of humanity cannot but wish that in all countries the labouring classes should have a taste for comforts and enjoyments, and that they should be stimulated by all legal means in their exertions to procure them. There cannot be a better security against a superabundant population. In those countries where the labouring classes have the fewest wants, and are contented with the cheapest food, the people are exposed to the greatest vicissitudes and miseries. They have no place of refuge from calamity; they cannot seek safety in a lower station; they are already so low that they can fall no lower. On any deficiency of the chief article of their subsistence there are few substitutes of which they can avail themselves and dearth to them is attended with almost all the evils of famine.

In the natural advance of society, the wages of labour will have a tendency to fall, as far as they are regulated by supply and demand; for the supply of labourers will continue to increase at the same rate, whilst the demand for them will increase at a slower rate. If, for instance, wages were regulated by a yearly increase of capital at the rate of 2 per cent., they would fall when it accumulated only at the rate of $1\frac{1}{2}$ per cent. They would fall still lower when it increased only at the rate of 1 or $\frac{1}{2}$ per cent., and would continue to do so until the capital became stationary, when wages also would become stationary, and be only sufficient to keep up the numbers of the actual population. I say that, under these circumstances, wages would fall if they were regulated only by the supply and demand of labourers; but we must not forget that wages are also regulated by the prices of the commodities on which they are expended.

As population increases, these necessaries will be constantly rising in price, because more labour will be necessary to produce them. If, then, the money wages of labour should fall, whilst every commodity on which the wages of labour were expended rose, the labourer would be doubly affected, and would be soon totally deprived of subsistence. Instead, therefore, of the money wages of labour falling, they would rise; but they would not rise sufficiently to enable the labourer to purchase as many comforts and necessaries as he did before the rise in the price of those commodities. If his annual wages were before £24, or six quarters of corn when the price was £4 per quarter, he would probably receive only the value of five quarters when corn rose to £5 per quarter. But five quarters would cost £25; he would, therefore, receive an addition in his money wages, though with that addition he would be unable to furnish himself with the same quantity of corn and other commodities which he had before consumed in his family.

Notwithstanding, then, that the labourer would be really worse paid, yet this increase in his wages would necessarily diminish the profits of the manufacturer; for his goods would sell at no higher price, and yet the expense of producing them would be increased. This, however, will be considered in our examination into the principles which regulate profits.

It appears, then, that the same cause which raises rent, namely, the increasing difficulty of providing an additional quantity of food with the same proportional quantity of labour, will also raise wages; and therefore, if money be of an unvarying value, both rent and wages will have a tendency to rise with the progress of wealth and population.

But there is this essential difference between the rise of rent and the rise of wages. The rise in the money value of rent is accompanied by an increased share of the produce; not only is the landlord's money rent greater, but his corn rent also; he will have more corn, and each defined measure of that corn will exchange for a greater quantity of all other goods which have not been raised in value. The fate of the labourer will be less happy; he will receive more money wages, it is true, but his corn wages will be reduced; and not only his command of corn, but his general condition will be deteriorated, by his finding it more difficult to maintain the market rate of wages above their natural rate. While the price of corn rises 10 per cent., wages will always rise less than 10 per cent., but rent will always rise more; the condition of the labourer will generally decline, and that of the landlord will always be improved.

When wheat was at £4 per quarter, suppose the labourer's wages to be £24 per annum, or the value of six quarters of wheat, and suppose half his wages to be expended on wheat, and the other half, or £12, on other things. He would receive

£24 14s.		£4 4s. 8d.		5.83 quarters.
£25 10s.	when wheat	£4 10s.	or the	5.66 quarters.
£26 8s.	was at	£4 16s.	value of	5.50 quarters.
£27 8s. 6d.		£5 2s. 10d.		5.33 quarters.

He would receive these wages to enable him to live just as well, and no better, than before; for when corn was at £4

per quarter, he would expend for three quarters of corn, at
£4 per quarter £12
and on other things £12
───
£24

When wheat was £4 4s. 8d., three quarters, which he and his
family consumed, would cost him . . . £12 14s.
other things not altered in price . . . £12
───
£24 14s.

When at £4 10s., three quarters of wheat would cost £13 10s.
and other things £12
───
£25 10s.

When at £4 16s., three quarters of wheat . . £14 8s.
other things £12
───
£26 8s.

When at £5 2s. 10d., three quarters of wheat would
cost . . . . £15 8s. 6d.
other things £12
───
£27 8s. 6d.

In proportion as corn became dear, he would receive less corn
wages, but his money wages would always increase, whilst his
enjoyments, on the above supposition, would be precisely the
same. But as other commodities would be raised in price in
proportion as raw produce entered into their composition, he
would have more to pay for some of them. Although his tea,
sugar, soap, candles, and house rent would probably be no
dearer, he would pay more for his bacon, cheese, butter, linen,
shoes, and cloth; and therefore, even with the above increase
of wages, his situation would be comparatively worse. But it
may be said that I have been considering the effect of wages on
price on the supposition that gold, or the metal from which
money is made, is the produce of the country in which wages
varied; and that the consequences which I have deduced agree
little with the actual state of things, because gold is a metal

of foreign production. The circumstance, however, of gold being a foreign production will not invalidate the truth of the argument, because it may be shown that whether it were found at home, or were imported from abroad, the effects ultimately, and, indeed, immediately, would be the same.

When wages rise it is generally because the increase of wealth and capital have occasioned a new demand for labour, which will infallibly be attended with an increased production of commodities. To circulate these additional commodities, even at the same prices as before, more money is required, more of this foreign commodity from which money is made, and which can only be obtained by importation. Whenever a commodity is required in greater abundance than before, its relative value rises comparatively with those commodities with which its purchase is made. If more hats were wanted, their price would rise, and more gold would be given for them. If more gold were required, gold would rise, and hats would fall in price, as a greater quantity of hats and of all other things would then be necessary to purchase the same quantity of gold. But in the case supposed, to say that commodities will rise because wages rise, is to affirm a positive contradiction; for we, first, say that gold will rise in relative value in consequence of demand, and, secondly, that it will fall in relative value because prices will rise, two effects which are totally incompatible with each other. To say that commodities are raised in price is the same thing as to say that money is lowered in relative value; for it is by commodities that the relative value of gold is estimated. If, then, all commodities rose in price, gold could not come from abroad to purchase those dear commodities, but it would go from home to be employed with advantage in purchasing the comparatively cheaper foreign commodities. It appears, then, that the rise of wages will not raise the prices of commodities, whether the metal from which money is made be produced at home or in a foreign country. All commodities cannot rise at the same time without an addition to the quantity of money. This addition could not be obtained at home, as we have already shown; nor could it be imported from abroad. To purchase any additional quantity of gold from abroad, commodities at home must be cheap, not dear. The importation of gold, and a rise in the price of all home-made commodities with which gold is purchased or paid for, are effects absolutely incompatible. The extensive use of paper money does not alter this question, for paper money conforms, or ought to conform, to the value of gold, and therefore its value is influenced by such causes only as influence the value of that metal.

These, then, are the laws by which wages are regulated, and by which the happiness of far the greatest part of every community is governed. Like all other contracts, wages should be left to the fair and free competition of the market, and should never be controlled by the interference of the legislature.

The clear and direct tendency of the poor laws is in direct opposition to these obvious principles: it is not, as the legislature benevolently intended, to amend the condition of the poor, but to deteriorate the condition of both poor and rich; instead of making the poor rich, they are calculated to make the rich poor; and whilst the present laws are in force, it is quite in the natural order of things that the fund for the maintenance of the poor should progressively increase till it has absorbed all the net revenue of the country, or at least so much of it as the state shall leave to us, after satisfying its own never-failing demands for the public expenditure.[1]

This pernicious tendency of these laws is no longer a mystery, since it has been fully developed by the able hand of Mr. Malthus; and every friend to the poor must ardently wish for their abolition. Unfortunately, however, they have been so long established, and the habits of the poor have been so formed upon their operation, that to eradicate them with safety from our political system requires the most cautious and skilful management. It is agreed by all who are most friendly to a repeal of these laws that, if it be desirable to prevent the most overwhelming distress to those for whose benefit they were erroneously enacted, their abolition should be effected by the most gradual steps.

It is a truth which admits not a doubt that the comforts and well-being of the poor cannot be permanently secured without some regard on their part, or some effort on the part of the legislature, to regulate the increase of their numbers, and to render less frequent among them early and improvident marriages. The operation of the system of poor laws has been directly contrary to this. They have rendered restraint super-

---

[1] With Mr. Buchanan, in the following passage, if it refers to temporary states of misery, I so far agree, that " the great evil of the labourer's condition is poverty, arising either from a scarcity of food or of work; and in all countries laws without number have been enacted for his relief. But there are miseries in the social state which legislation cannot relieve; and it is useful therefore to know its limits, that we may not, by aiming at what is impracticable, miss the good which is really in our power."— Buchanan, p. 61.

fluous, and have invited imprudence, by offering it a portion of the wages of prudence and industry.[1]

The nature of the evil points out the remedy. By gradually contracting the sphere of the poor laws; by impressing on the poor the value of independence, by teaching them that they must look not to systematic or casual charity, but to their own exertions for support, that prudence and forethought are neither unnecessary nor unprofitable virtues, we shall by degrees approach a sounder and more healthful state.

No scheme for the amendment of the poor laws merits the least attention which has not their abolition for its ultimate object; and he is the best friend of the poor, and to the cause of humanity, who can point out how this end can be attained with the most security, and at the same time with the least violence. It is not by raising in any manner different from the present the fund from which the poor are supported that the evil can be mitigated. It would not only be no improvement, but it would be an aggravation of the distress which we wish to see removed, if the fund were increased in amount or were levied according to some late proposals, as a general fund from the country at large. The present mode of its collection and application has served to mitigate its pernicious effects. Each parish raises a separate fund for the support of its own poor. Hence it becomes an object of more interest and more practicability to keep the rates low than if one general fund were raised for the relief of the poor of the whole kingdom. A parish is much more interested in an economical collection of the rate, and a sparing distribution of relief, when the whole saving will be for its own benefit, than if hundreds of other parishes were to partake of it.

It is to this cause that we must ascribe the fact of the poor laws not having yet absorbed all the net revenue of the country; it is to the rigour with which they are applied that we are indebted for their not having become overwhelmingly oppres-

---

[1] The progress of knowledge manifested upon this subject in the House of Commons since 1796 has happily not been very small, as may be seen by contrasting the late report of the committee on the poor laws and the following sentiments of Mr. Pitt in that year: " Let us," said he, " make relief in cases where there are a number of children a matter of right and honour, instead of a ground of opprobrium and contempt. This will make a large family a blessing and not a curse; and this will draw a proper line of distinction between those who are able to provide for themselves by their labour, and those who, after having enriched their country with a number of children, have a claim upon its assistance for support."— Hansard's *Parliamentary History*, vol. xxxii. p. 710.

sive. If by law every human being wanting support could be sure to obtain it, and obtain it in such a degree as to make life tolerably comfortable, theory would lead us to expect that all other taxes together would be light compared with the single one of poor rates. The principle of gravitation is not more certain than the tendency of such laws to change wealth and power into misery and weakness; to call away the exertions of labour from every object, except that of providing mere subsistence; to confound all intellectual distinction; to busy the mind continually in supplying the body's wants; until at last all classes should be infected with the plague of universal poverty. Happily these laws have been in operation during a period of progressive prosperity, when the funds for the maintenance of labour have regularly increased, and when an increase of population would be naturally called for. But if our progress should become more slow; if we should attain the stationary state, from which I trust we are yet far distant, then will the pernicious nature of these laws become more manifest and alarming; and then, too, will their removal be obstructed by many additional difficulties.

CHAPTER VI

ON PROFITS

THE profits of stock, in different employments, having been shown to bear a proportion to each other, and to have a tendency to vary all in the same degree and in the same direction, it remains for us to consider what is the cause of the permanent variations in the rate of profit, and the consequent permanent alterations in the rate of interest.

We have seen that the price[1] of corn is regulated by the quantity of labour necessary to produce it, with that portion of capital which pays no rent. We have seen, too, that all manufactured commodities rise and fall in price in proportion as more or less labour becomes necessary to their production. Neither the farmer who cultivates that quantity of land which regulates price, nor the manufacturer who manufactures goods, sacrifice any portion of the produce for rent. The whole value of their commodities is divided into two portions only: one constitutes the profits of stock, the other the wages of labour.

Supposing corn and manufactured goods always to sell at the same price, profits would be high or low in proportion as wages were low or high. But suppose corn to rise in price because more labour is necessary to produce it; that cause will not raise the price of manufactured goods in the production of which no additional quantity of labour is required. If, then, wages continued the same, the profits of manufacturers would remain the same; but if, as is absolutely certain, wages should rise with the rise of corn, then their profits would necessarily fall.

If a manufacturer always sold his goods for the same money, for £1000, for example, his profits would depend on the price of the labour necessary to manufacture those goods. His profits would be less when wages amounted to £800 than when he paid only £600. In proportion then as wages rose would profits fall. But if the price of raw produce would increase, it may be asked

[1] The reader is desired to bear in mind that, for the purpose of making the subject more clear, I consider money to be invariable in value, and therefore every variation of price to be referable to an alteration in the value of the commodity.

whether the farmer at least would not have the same rate of profits, although he should pay an additional sum for wages? Certainly not: for he will not only have to pay, in common with the manufacturer, an increase of wages to each labourer he employs, but he will be obliged either to pay rent, or to employ an additional number of labourers to obtain the same produce; and the rise in the price of raw produce will be proportioned only to that rent, or that additional number, and will not compensate him for the rise of wages.

If both the manufacturer and farmer employed ten men, on wages rising from £24 to £25 per annum per man, the whole sum paid by each would be £250 instead of £240. This is, however, the whole addition that would be paid by the manufacturer to obtain the same quantity of commodities; but the farmer on new land would probably be obliged to employ an additional man, and therefore to pay an additional sum of £25 for wages; and the farmer on the old land would be obliged to pay precisely the same additional sum of £25 for rent; without which additional labour corn would not have risen nor rent have been increased. One will therefore have to pay £275 for wages alone, the other for wages and rent together; each £25 more than the manufacturer: for this latter £25 the farmer is compensated by the addition to the price of raw produce, and therefore his profits still conform to the profits of the manufacturer. As this proposition is important, I will endeavour still further to elucidate it.

We have shown that in early stages of society, both the landlord's and the labourer's share of the *value* of the produce of the earth would be but small; and that it would increase in proportion to the progress of wealth and the difficulty of procuring food. We have shown, too, that although the value of the labourer's portion will be increased by the high value of food, his real share will be diminished; whilst that of the landlord will not only be raised in value, but will also be increased in quantity.

The remaining quantity of the produce of the land, after the landlord and labourer are paid, necessarily belongs to the farmer, and constitutes the profits of his stock. But it may be alleged, that though, as society advances, his proportion of the whole produce will be diminished, yet as it will rise in value, he, as well as the landlord and labourer, may, notwithstanding, receive a greater value.

It may be said, for example, that when corn rose from £4 to

£10, the 180 quarters obtained from the best land would sell for £1800 instead of £720; and, therefore, though the landlord and labourer be proved to have a greater value for rent and wages, still the value of the farmer's profit might also be augmented. This, however, is impossible, as I shall now endeavour to show.

In the first place, the price of corn would rise only in proportion to the increased difficulty of growing it on land of a worse quality.

It has been already remarked, that if the labour of ten men will, on land of a certain quality, obtain 180 quarters of wheat, and its value be £4 per quarter, or £720; and if the labour of ten additional men will, on the same or any other land, produce only 170 quarters in addition, wheat would rise from £4 to £4 4s. 8d.; for 170 : 180 : : £4 : £4 4s. 8d. In other words, as for the production of 170 quarters the labour of ten men is necessary in the one case, and only that of 9.44 in the other, the rise would be as 9.44 to 10, or as £4 to £4 4s. 8d. In the same manner it might be shown that, if the labour of ten additional men would only produce 160 quarters, the price would further rise to £4 10s.; if 150, to £4 16s., etc., etc.

But when 180 quarters were produced on the land paying
no rent, and its price was £4 per quarter, it is sold
for £720
And when 170 quarters were produced on the land paying
no rent, and the price rose to £4 4s. 8d., it still sold
for 720
So 160 quarters at £4 10s. produce 720
And 150 quarters at £4 16s. produce the same sum of . 720

Now, it is evident that if, out of these equal values, the farmer is at one time obliged to pay wages regulated by the price of wheat at £4, and at other times at higher prices, the rate of his profits will diminish in proportion to the rise in the price of corn.

In this case, therefore, I think it is clearly demonstrated that a rise in the price of corn, which increases the money wages of the labourer, diminishes the money value of the farmer's profits.

But the case of the farmer of the old and better land will be in no way different; he also will have increased wages to pay, and will never retain more of the value of the produce, however high may be its price, than £720 to be divided between himself and his always equal number of labourers; in proportion therefore as they get more, he must retain less.

When the price of corn was at £4, the whole 180 quarters

belonged to the cultivator, and he sold it for £720. When corn rose to £4 4s. 8d., he was obliged to pay the value of ten quarters out of his 180 for rent, consequently the remaining 170 yielded him no more than £720: when it rose further to £4 10s., he paid twenty quarters, or their value, for rent, and consequently only retained 160 quarters, which yielded the same sum of £720.

It will be seen, then, that whatever rise may take place in the price of corn, in consequence of the necessity of employing more labour and capital to obtain a given additional quantity of produce, such rise will always be equalled in value by the additional rent or additional labour employed; so that whether corn sells for £4, £4 10s., or £5 2s. 10d., the farmer will obtain for that which remains to him, after paying rent, the same real value. Thus we see that whether the produce belonging to the farmer be 180, 170, 160, or 150 quarters, he always obtains the same sum of £720 for it; the price increasing in an inverse proportion to the quantity.

Rent, then, it appears, always falls on the consumer, and never on the farmer; for if the produce of his farm should uniformly be 180 quarters, with the rise of price he would retain the value of a less quantity for himself, and give the value of a larger quantity to his landlord; but the deduction would be such as to leave him always the same sum of £720.

It will be seen too, that, in all cases, the same sum of £720 must be divided between wages and profits. If the value of the raw produce from the land exceed this value it belongs to rent, whatever may be its amount. If there be no excess, there will be no rent. Whether wages or profits rise or fall, it is this sum of £720 from which they must both be provided. On the one hand, profits can never rise so high as to absorb so much of this £720 that enough will not be left to furnish the labourers with absolute necessaries; on the other hand, wages can never rise so high as to leave no portion of this sum for profits.

Thus in every case, agricultural as well as manufacturing profits are lowered by a rise in the price of raw produce, if it be accompanied by a rise of wages.[1] If the farmer gets no additional value for the corn which remains to him after paying rent, if the manufacturer gets no additional value for the goods which he manufactures, and if both are obliged to pay a greater

---

[1] The reader is aware that we are leaving out of our consideration the accidental variations arising from bad and good seasons, or from the demand increasing or diminishing by any sudden effect on the state of population. We are speaking of the natural and constant, not of the accidental and fluctuating, price of corn.

value in wages, can any point be more clearly established than that profits must fall with a rise of wages?

The farmer, then, although he pays no part of his landlord's rent, that being always regulated by the price of produce, and invariably falling on the consumers, has however a very decided interest in keeping rent low, or rather in keeping the natural price of produce low. As a consumer of raw produce, and of those things into which raw produce enters as a component part, he will, in common with all other consumers, be interested in keeping the price low. But he is most materially concerned with the high price of corn as it affects wages. With every rise in the price of corn, he will have to pay, out of an equal and unvarying sum of £720, an additional sum for wages to the ten men whom he is supposed constantly to employ. We have seen, in treating on wages, that they invariably rise with the rise in the price of raw produce. On a basis assumed for the purpose of calculation, page 58, it will be seen that if when wheat is at £4 per quarter, wages should be £24 per annum,

	£	s.	d.			£	s.	d.
When wheat is at	4	4	8	wages would be		24	14	0
	4	10	0			25	10	0
	4	16	0			26	8	0
	5	2	10			27	8	6

Now, of the unvarying fund of £720 to be distributed between labourers and farmers,

	£	s.	d.		£	s.	d.		£	s.	d.
When the price of wheat is at	4	0	0	the labourers will receive	240	0	0	the farmer will receive	480	0	0
	4	4	8		247	0	0		473	0	0
	4	10	0		255	0	0		465	0	0
	4	16	0		264	0	0		456	0	0
	5	2	10		274	5	0		445	15	1

[1] The 180 quarters of corn would be divided in the following proportions between landlords, farmers, and labourers, with the above-named variations in the value of corn.

Price per qr.	Rent.	Profit.	Wages.	Total.
£ s. d.	In Wheat.	In Wheat.	In Wheat.	
4 0 0	None.	120 qrs.	60 qrs.	
4 4 8	10 qrs.	111.7	58.3	
4 10 0	20	103.4	56.6	180
4 16 0	30	95	55	
5 2 10	40	86.7	53.3	

and, under the same circumstances, money rent, wages, and profit would be as follows:

Price per qr.	Rent.			Profit.			Wages.			Total.		
£ s. d.	£	s.	d.	£	s.	d.	£	s.	d.	£	s.	d.
4 0 0	None.			480	0	0	240	0	0	720	0	0
4 4 8	42	7	6	473	0	0	247	0	0	762	7	6
4 10 0	90	0	0	465	0	0	255	0	0	810	0	0
4 16 0	144	0	0	456	0	0	264	0	0	864	0	0
5 2 10	205	13	4	445	15	0	274	5	0	925	13	4

And supposing that the original capital of the farmer was £3000, the profits of his stock being in the first instance £480, would be at the rate of 16 per cent. When his profits fell to £473, they would be at the rate of 15.7 per cent.

£465	.	.	.	.	15.5
£456	.	.	.	.	15.2
£445	.	.	.	14.8

But the *rate* of profits will fall still more, because the capital of the farmer, it must be recollected, consists in a great measure of raw produce, such as his corn and hay-ricks, his unthreshed wheat and barley, his horses and cows, which would all rise in price in consequence of the rise of produce. His absolute profits would fall from £480 to £445 15s.; but if, from the cause which I have just stated, his capital should rise from £3000 to £3200, the rate of his profits would, when corn was at £5 2s. 10d., be under 14 per cent.

If a manufacturer had also employed £3000 in his business, he would be obliged, in consequence of the rise of wages, to increase his capital, in order to be enabled to carry on the same business. If his commodities sold before for £720 they would continue to sell at the same price; but the wages of labour, which were before £240, would rise, when corn was at £5 2s. 10d., to £274 5s. In the first case he would have a balance of £480 as profit on £3000, in the second he would have a profit only of £445 15s., on an increased capital, and therefore his profits would conform to the altered rate of those of the farmer.

There are few commodities which are not more or less affected in their price by the rise of raw produce, because some raw material from the land enters into the composition of most commodities. Cotton goods, linen, and cloth will all rise in price with the rise of wheat; but they rise on account of the greater quantity of labour expended on the raw material from which they are made, and not because more was paid by the manufacturer to the labourers whom he employed on those commodities.

In all cases, commodities rise because more labour is expended on them, and not because the labour which is expended on them is at a higher value. Articles of jewellery, of iron, of plate, and of copper, would nor rise, because none of the raw produce from the surface of the earth enters into their composition.

It may be said that I have taken it for granted that money wages would rise with a rise in the price of raw produce, but that this is by no means a necessary consequence, as the labourer may be contented with fewer enjoyments. It is true that the wages of labour may previously have been at a high level, and that they may bear some reduction. If so, the fall of profits will be checked; but it is impossible to conceive that the money price of wages should fall or remain stationary with a gradually increasing price of necessaries; and therefore it may be taken for granted that, under ordinary circumstances, no permanent rise takes place in the price of necessaries without occasioning, or having been preceded by, a rise in wages.

The effects produced on profits would have been the same, or nearly the same, if there had been any rise in the price of those other necessaries, besides food, on which the wages of labour are expended. The necessity which the labourer would be under of paying an increased price for such necessaries would oblige him to demand more wages; and whatever increases wages, necessarily reduces profits. But suppose the price of silks, velvets, furniture, and any other commodities, not required by the labourer, to rise in consequence of more labour being expended on them, would not that affect profits? Certainly not: for nothing can affect profits but a rise in wages; silks and velvets are not consumed by the labourer, and therefore cannot raise wages.

It is to be understood that I am speaking of profits generally. I have already remarked that the market price of a commodity may exceed its natural or necessary price, as it may be produced in less abundance than the new demand for it requires. This, however, is but a temporary effect. The high profits on capital employed in producing that commodity will naturally attract capital to that trade; and as soon as the requisite funds are supplied, and the quantity of the commodity is duly increased, its price will fall, and the profits of the trade will conform to the general level. A fall in the general rate of profits is by no means incompatible with a partial rise of profits in particular employments. It is through the inequality of profits that capital is moved from one employment to another. Whilst, then, general profits are falling, and gradually settling at a lower level in consequence of the rise of wages, and the increasing difficulty of supplying the increasing population with necessaries, the profits of the farmer may, for an interval of some little duration, be above the former level. An extraordinary stimulus may be also given for a certain time to a particular branch of foreign and colonial trade; but the admission of this fact by no means invalidates the theory, that profits depend on high or low wages, wages on the price of necessaries, and the price of necessaries chiefly on the price of food, because all other requisites may be increased almost without limit.

It should be recollected that prices always vary in the market, and in the first instance, through the comparative state of demand and supply. Although cloth could be furnished at 40s. per yard, and give the usual profits of stock, it may rise to 60s. or 80s. from a general change of fashion, or from any other cause which should suddenly and unexpectedly increase the demand or diminish the supply of it. The makers of cloth will for a time have unusual profits, but capital will naturally flow to that manufacture, till the supply and demand are again at their fair level, when the price of cloth will again sink to 40s., its natural or necessary price. In the same manner, with every increased demand for corn, it may rise so high as to afford more than the general profits to the farmer. If there be plenty of fertile land, the price of corn will again fall to its former standard, after the requisite quantity of capital has been employed in producing it, and profits will be as before; but if there be not plenty of fertile land, if, to produce this additional quantity, more than the usual quantity of capital and labour be required, corn will not fall to its former level. Its natural price will be raised, and the farmer, instead of obtaining permanently larger profits, will find himself obliged to be satisfied with the diminished rate which is the inevitable consequence of the rise of wages, produced by the rise of necessaries.

The natural tendency of profits then is to fall; for, in the progress of society and wealth, the additional quantity of food required is obtained by the sacrifice of more and more labour. This tendency, this gravitation as it were of profits, is happily checked at repeated intervals by the improvements in machinery connected with the production of necessaries, as well as by discoveries in the science of agriculture, which enable us to relinquish a portion of labour before required, and therefore to lower the price of the prime necessary of the labourer. The rise in the price of necessaries and in the wages of labour is, however, limited; for as soon as wages should be equal (as in the case formerly stated) to £720, the whole receipts of the farmer, there must be an end of accumulation; for no capital can then yield any profit whatever, and no additional labour can be demanded, and consequently population will have reached its highest point. Long, indeed, before this period, the very low rate of profits will have arrested all accumulation, and almost the whole produce of the country, after paying the labourers, will be the property of the owners of land and the receivers of tithes and taxes.

Thus, taking the former very imperfect basis as the grounds of my calculation, it would appear that when corn was at £20 per quarter, the whole net income of the country would belong to the landlords, for then the same quantity of labour that was originally necessary to produce 180 quarters would be necessary to produce 36; since £20 : £4 : : 180 : 36. The farmer, then, who produced 180 quarters (if any such there were, for the old and new capital employed on the land would be so blended that it could in no way be distinguished), would sell the

180 qrs. at £20 per qr. or		£3600
the value of 144 qrs. { to landlord for rent, being the difference	between 36 and 180 qrs. }	2880
36 qrs.		720
the value of 36 qrs. to labourers, ten in number . . .		720

leaving nothing whatever for profit.

I have supposed that at this price of £20 the labourers would continue to consume three quarters each per annum, or   £60
And that on the other commodities they would
expend   12
                                     72 for each labourer.
And therefore ten labourers would cost £720 per annum.

In all these calculations I have been desirous only to elucidate the principle, and it is scarcely necessary to observe that my whole basis is assumed at random, and merely for the purpose of exemplification. The results, though different in degree, would have been the same in principle, however accurately I might have set out in stating the difference in the number of labourers necessary to obtain the successive quantities of corn required by an increasing population, the quantity consumed by the labourer's family, etc., etc. My object has been to simplify the subject, and I have therefore made no allowance for the increasing price of the other necessaries, besides food, of the labourer; an increase which would be the consequence of the increased value of the raw materials from which they are made, and which would of course further increase wages and lower profits.

I have already said that long before this state of prices was become permanent there would be no motive for accumulation;

for no one accumulates but with a view to make his accumulation productive, and it is only when so employed that it operates on profits. Without a motive there could be no accumulation, and consequently such a state of prices never could take place. The farmer and manufacturer can no more live without profit than the labourer without wages. Their motive for accumulation will diminish with every diminution of profit, and will cease altogether when their profits are so low as not to afford them an adequate compensation for their trouble, and the risk which they must necessarily encounter in employing their capital productively.

I must again observe that the rate of profits would fall much more rapidly than I have estimated in my calculation; for the value of the produce being what I have stated it under the circumstances supposed, the value of the farmer's stock would be greatly increased from its necessarily consisting of many of the commodities which had risen in value. Before corn could rise from £4 to £12, his capital would probably be doubled in exchangeable value, and be worth £6000 instead of £3000. If then his profit were £180, or 6 per cent. on his original capital, profits would not at that time be really at a higher *rate* than 3 per cent.; for £6000 at 3 per cent. gives £180; and on those terms only could a new farmer with £6000 money in his pocket enter into the farming business.

Many trades would derive some advantage, more or less, from the same source. The brewer, the distiller, the clothier, the linen manufacturer, would be partly compensated for the diminution of their profits by the rise in the value of their stock of raw and finished materials; but a manufacturer of hardware, of jewellery, and of many other commodities, as well as those whose capitals uniformly consisted of money, would be subject to the whole fall in the rate of profits, without any compensation whatever.

We should also expect that, however the rate of the profits of stock might diminish in consequence of the accumulation of capital on the land, and the rise of wages, yet that the aggregate amount of profits would increase. Thus, supposing that, with repeated accumulations of £100,000, the rate of profit should fall from 20 to 19, to 18, to 17 per cent., a constantly diminishing rate, we should expect that the whole amount of profits received by those successive owners of capital would be always progressive; that it would be greater when the capital was £200,000 than when £100,000; still greater when £300,000; and so on, increasing, though at a diminishing rate, with every increase of capital. This progression, however, is only true for a certain time; thus, 19 per cent. on £200,000 is more than 20 on £100,000; again, 18 per cent. on £300,000 is more than 19 per cent. on £200,000; but after capital has accumulated to a large amount, and profits have fallen, the further accumulation diminishes the aggregate of profits. Thus, suppose the accumulation should be £1,000,000, and the profits 7 per cent., the whole amount of profits will be £70,000; now if an addition of £100,000 capital be made to the million, and profits should fall to 6 per cent., £66,000 or a diminution of £4000 will be received by the owners of stock, although the whole amount of stock will be increased from £1,000,000 to £1,100,000.

There can, however, be no accumulation of capital so long as stock yields any profit at all, without its yielding not only an increase of produce, but an increase of value. By employing £100,000 additional capital, no part of the former capital will be rendered less productive. The produce of the land and labour of the country must increase, and its value will be raised, not only by the value of the addition which is made to the former quantity of productions, but by the new value which is given to the whole produce of the land, by the increased difficulty of producing the last portion of it. When the accumulation of capital, however, becomes very great, notwithstanding this increased value, it will be so distributed that a less value than before will be appropriated to profits, while that which is devoted to rent and wages will be increased. Thus with successive additions of £100,000 to capital, with a fall in the rate of profits, from 20 to 19, to 18, to 17 per cent., etc., the productions annually obtained will increase in quantity, and be of more than the whole additional value which the additional capital is calculated to produce. From £20,000 it will rise to more than £39,000, and then to more than £57,000, and when the capital employed is a million, as we before supposed, if £100,000 more be added to it, and the aggregate of profits is actually lower than before, more than £6000 will nevertheless be added to the revenue of the country, but it will be to the revenue of the landlords and labourers; they will obtain more than the additional produce, and will from their situation be enabled to encroach even on the former gains of the capitalist. Thus, suppose the price of corn to be £4 per quarter, and that therefore, as we before calculated, of every £720 remaining to the farmer after payment of his rent, £480 were retained by him, and £240 were paid to his labourers; when the price rose to £6 per quarter, he would be obliged to pay his labourers £300 and retain only £420 for profits: he would be obliged to pay them £300 to enable them to consume the same quantity of necessaries as before, and no more. Now if the capital employed were so large as to yield a hundred thousand times £720, or £72,000,000, the aggregate of profits would be £48,000,000 when wheat was at £4 per quarter; and if by employing a larger capital 105,000 times £720 were obtained when wheat was at £6, or £75,600,000, profits would actually fall from £48,000,000 to £44,100,000 or 105,000 times £420, and wages would rise from £24,000,000 to £31,500,000. Wages would rise because more labourers would be employed in proportion to capital; and each labourer would receive more money wages; but the condition of the labourer, as we have already shown, would be worse, inasmuch as he would be able to command a less quantity of the produce of the country. The only real gainers would be the landlords; they would receive higher rents, first, because produce would be of a higher value, and secondly, because they would have a greatly increased proportion of that produce.

Although a greater value is produced, a greater proportion of what remains of that value, after paying rent, is consumed by the producers, and it is this, and this alone, which regulates profits. Whilst the land yields abundantly, wages may temporarily rise, and the producers may consume more than their accustomed proportion; but the stimulus which will thus be given to population will speedily reduce the labourers to their usual consumption. But when poor lands are taken into cultivation, or when more capital and labour are expended on the old land, with a less return of produce, the effect must be permanent. A greater proportion of that part of the produce which remains to be divided, after paying rent, between the owners of stock and the labourers, will be apportioned to the latter. Each man may, and probably will, have a less absolute quantity; but as more labourers are employed in proportion to the whole produce retained by the farmer, the value of a greater proportion of the whole produce will be absorbed by wages, and consequently the value of a smaller proportion will be devoted to profits. This will necessarily be rendered permanent by the laws of nature, which have limited the productive powers of the land.

Thus we again arrive at the same conclusion which we have

before attempted to establish:—that in all countries, and all times, profits depend on the quantity of labour requisite to provide necessaries for the labourers on that land or with that capital which yields no rent. The effects then of accumulation will be different in different countries, and will depend chiefly on the fertility of the land. However extensive a country may be where the land is of a poor quality, and where the importation of food is prohibited, the most moderate accumulations of capital will be attended with great reductions in the rate of profit and a rapid rise in rent; and on the contrary a small but fertile country, particularly if it freely permits the importation of food, may accumulate a large stock of capital without any great diminution in the rate of profits, or any great increase in the rent of land. In the Chapter on Wages we have endeavoured to show that the money price of commodities would not be raised by a rise of wages, either on the supposition that gold, the standard of money, was the produce of this country, or that it was imported from abroad. But if it were otherwise, if the prices of commodities were permanently raised by high wages, the proposition would not be less true, which asserts that high wages invariably affect the employers of labour by depriving them of a portion of their real profits. Supposing the hatter, the hosier, and the shoemaker each paid £10 more wages in the manufacture of a particular quantity of their commodities, and that the price of hats, stockings, and shoes rose by a sum sufficient to repay the manufacturer the £10; their situation would be no better than if no such rise took place. If the hosier sold his stockings for £110 instead of £100, his profits would be precisely the same money amount as before; but as he would obtain in exchange for this equal sum, one-tenth less of hats, shoes, and every other commodity, and as he could with his former amount of savings employ fewer labourers at the increased wages, and purchase fewer raw materials at the increased prices, he would be in no better situation than if his money profits had been really diminished in amount and everything had remained at its former price. Thus, then, I have endeavoured to show, first, that a rise of wages would not raise the price of commodities, but would invariably lower profits; and secondly, that if the prices of all commodities could be raised, still the effect on profits would be the same; and that, in fact, the value of the medium only in which prices and profits are estimated would be lowered.

CHAPTER VII

ON FOREIGN TRADE

No extension of foreign trade will immediately increase the amount of value in a country, although it will very powerfully contribute to increase the mass of commodities, and therefore the sum of enjoyments. As the value of all foreign goods is measured by the quantity of the produce of our land and labour which is given in exchange for them, we should have no greater value if, by the discovery of new markets, we obtained double the quantity of foreign goods in exchange for a given quantity of ours. If by the purchase of English goods to the amount of £1000 a merchant can obtain a quantity of foreign goods, which he can sell in the English market for £1200, he will obtain 20 per cent. profit by such an employment of his capital; but neither his gains, nor the value of the commodities imported, will be increased or diminished by the greater or smaller quantity of foreign goods obtained. Whether, for example, he imports twenty-five or fifty pipes of wine, his interest can be no way affected if at one time the twenty-five pipes, and at another the fifty pipes, equally sell for £1200. In either case his profit will be limited to £200, or 20 per cent. on his capital; and in either case the same value will be imported into England. If the fifty pipes sold for more than £1200, the profits of this individual merchant would exceed the general rate of profits, and capital would naturally flow into this advantageous trade, till the fall of the price of wine had brought everything to the former level.

It has indeed been contended that the great profits which are sometimes made by particular merchants in foreign trade will elevate the general rate of profits in the country, and that the abstraction of capital from other employments, to partake of the new and beneficial foreign commerce, will raise prices generally, and thereby increase profits. It has been said, by high authority, that less capital being necessarily devoted to the growth of corn, to the manufacture of cloth, hats, shoes, etc., while the demand continues the same, the price of these commodities will be so increased, that the farmer, hatter, clothier, and shoemaker will have an increase of profits as well as the foreign merchant.[1]

They who hold this argument agree with me that the profits of different employments have a tendency to conform to one another; to advance and recede together. Our variance consists in this: They contend that the equality of profits will be brought about by the general rise of profits; and I am of opinion that the profits of the favoured trade will speedily subside to the general level.

For, first, I deny that less capital will necessarily be devoted to the growth of corn, to the manufacture of cloth, hats, shoes, etc., unless the demand for these commodities be diminished; and if so, their price will not rise. In the purchase of foreign commodities, either the same, a larger, or a less portion of the produce of the land and labour of England will be employed. If the same portion be so employed, then will the same demand exist for cloth, shoes, corn, and hats as before, and the same portion of capital will be devoted to their production. If, in consequence of the price of foreign commodities being cheaper, a less portion of the annual produce of the land and labour of England is employed in the purchase of foreign commodities, more will remain for the purchase of other things. If there be a greater demand for hats, shoes, corn, etc., than before, which there may be, the consumers of foreign commodities having an additional portion of their revenue disposable, the capital is also disposable with which the greater value of foreign commodities was before purchased; so that with the increased demand for corn, shoes, etc., there exists also the means of procuring an increased supply, and therefore neither prices nor profits can permanently rise. If more of the produce of the land and labour of England be employed in the purchase of foreign commodities, less can be employed in the purchase of other things, and therefore fewer hats, shoes, etc., will be required. At the same time that capital is liberated from the production of shoes, hats, etc., more must be employed in manufacturing those commodities with which foreign commodities are purchased; and, consequently, in all cases the demand for foreign and home commodities together, as far as regards value, is limited by the revenue and capital of the country. If one increases the other must diminish. If the quantity of wine imported in exchange for the same quantity of English com-

[1] See Adam Smith, book i. chap. 9.

modities be doubled, the people of England can either consume double the quantity of wine that they did before, or the same quantity of wine and a greater quantity of English commodities. If my revenue had been £1000 with which I purchased annually one pipe of wine for £100, and a certain quantity of English commodities for £900; when wine fell to £50 per pipe, I might lay out the £50 saved, either in the purchase of an additional pipe of wine or in the purchase of more English commodities. If I bought more wine, and every wine-drinker did the same, the foreign trade would not be in the least disturbed; the same quantity of English commodities would be exported in exchange for wine, and we should receive double the quantity, though not double the value of wine. But if I, and others, contented ourselves with the same quantity of wine as before, fewer English commodities would be exported, and the wine-drinkers might either consume the commodities which were before exported, or any others for which they had an inclination. The capital required for their production would be supplied by the capital liberated from the foreign trade.

There are two ways in which capital may be accumulated; it may be saved either in consequence of increased revenue or of diminished consumption. If my profits are raised from £1000 to £1200, while my expenditure continues the same, I accumulate annually £200 more than I did before. If I save £200 out of my expenditure, while my profits continue the same, the same effect will be produced; £200 per annum will be added to my capital. The merchant who imported wine after profits had been raised from 20 per cent. to 40 per cent., instead of purchasing his English goods for £1000, must purchase them for £857 2s. 10d., still selling the wine which he imports in return for those goods for £1200; or, if he continued to purchase his English goods for £1000, must raise the price of his wine to £1400; he would thus obtain 40 instead of 20 per cent. profit on his capital; but if, in consequence of the cheapness of all the commodities on which his revenue was expended, he and all other consumers could save the value of £200 out of every £1000 they before expended, they would more effectually add to the real wealth of the country; in one case, the savings would be made in consequence of an increase of revenue, in the other, in consequence of diminished expenditure.

If, by the introduction of machinery, the generality of the commodities on which revenue was expended fell 20 per cent. in value, I should be enabled to save as effectually as if my revenue had been raised 20 per cent.; but in one case the rate of profits is stationary, in the other it is raised 20 per cent.—If, by the introduction of cheap foreign goods, I can save 20 per cent. from my expenditure, the effect will be precisely the same as if machinery had lowered the expense of their production, but profits would not be raised.

It is not, therefore, in consequence of the extension of the market that the rate of profit is raised, although such extension may be equally efficacious in increasing the mass of commodities, and may thereby enable us to augment the funds destined for the maintenance of labour, and the materials on which labour may be employed. It is quite as important to the happiness of mankind that our enjoyments should be increased by the better distribution of labour, by each country producing those commodities for which by its situation, its climate, and its other natural or artificial advantages it is adapted, and by their exchanging them for the commodities of other countries, as that they should be augmented by a rise in the rate of profits.

It has been my endeavour to show throughout this work that the rate of profits can never be increased but by a fall in wages, and that there can be no permanent fall of wages but in consequence of a fall of the necessaries on which wages are expended. If, therefore, by the extension of foreign trade, or by improvements in machinery, the food and necessaries of the labourer can be brought to market, at a reduced price, profits will rise. If, instead of growing our own corn, or manufacturing the clothing and other necessaries of the labourer, we discover a new market from which we can supply ourselves with these commodities at a cheaper price, wages will fall and profits rise; but if the commodities obtained at a cheaper rate, by the extension of foreign commerce, or by the improvement of machinery, be exclusively the commodities consumed by the rich, no alteration will take place in the rate of profits. The rate of wages would not be affected, although wine, velvets, silks, and other expensive commodities should fall 50 per cent., and consequently profits would continue unaltered.

Foreign trade, then, though highly beneficial to a country, as it increases the amount and variety of the objects on which revenue may be expended, and affords, by the abundance and cheapness of commodities, incentives to saving, and to the accumulation of capital, has no tendency to raise the profits of stock unless the commodities imported be of that description on which the wages of labour are expended.

The remarks which have been made respecting foreign trade apply equally to home trade. The rate of profits is never increased by a better distribution of labour, by the invention of machinery, by the establishment of roads and canals, or by any means of abridging labour either in the manufacture or in the conveyance of goods. These are causes which operate on price, and never fail to be highly beneficial to consumers; since they enable them, with the same labour, or with the value of the produce of the same labour, to obtain in exchange a greater quantity of the commodity to which the improvement is applied; but they have no effect whatever on profit. On the other hand, every diminution in the wages of labour raises profits, but produces no effect on the price of commodities. One is advantageous to all classes, for all classes are consumers; the other is beneficial only to producers; they gain more, but everything remains at its former price. In the first case they get the same as before; but everything on which their gains are expended is diminished in exchangeable value.

The same rule which regulates the relative value of commodities in one country does not regulate the relative value of the commodities exchanged between two or more countries.

Under a system of perfectly free commerce, each country naturally devotes its capital and labour to such employments as are most beneficial to each. This pursuit of individual advantage is admirably connected with the universal good of the whole. By stimulating industry, by rewarding ingenuity, and by using most efficaciously the peculiar powers bestowed by nature, it distributes labour most effectively and most economically: while, by increasing the general mass of productions, it diffuses general benefit, and binds together, by one common tie of interest and intercourse, the universal society of nations throughout the civilised world. It is this principle which determines that wine shall be made in France and Portugal, that corn shall be grown in America and Poland, and that hardware and other goods shall be manufactured in England.

In one and the same country, profits are, generally speaking, always on the same level; or differ only as the employment of capital may be more or less secure and agreeable. It is not so between different countries. If the profits of capital employed in Yorkshire should exceed those of capital employed in London, capital would speedily move from London to Yorkshire, and an equality of profits would be effected; but if in consequence of the diminished rate of production in the lands of England from the increase of capital and population wages should rise and profits fall, it would not follow that capital and population would necessarily move from England to Holland, or Spain, or Russia, where profits might be higher.

If Portugal had no commercial connection with other countries, instead of employing a great part of her capital and industry in the production of wines, with which she purchases for her own use the cloth and hardware of other countries, she would be obliged to devote a part of that capital to the manufacture of those commodities, which she would thus obtain probably inferior in quality as well as quantity.

The quantity of wine which she shall give in exchange for the cloth of England is not determined by the respective quantities of labour devoted to the production of each, as it would be if both commodities were manufactured in England, or both in Portugal.

England may be so circumstanced that to produce the cloth may require the labour of 100 men for one year; and if she attempted to make the wine, it might require the labour of 120 men for the same time. England would therefore find it her interest to import wine, and to purchase it by the exportation of cloth.

To produce the wine in Portugal might require only the labour of 80 men for one year, and to produce the cloth in the same country might require the labour of 90 men for the same time. It would therefore be advantageous for her to export wine in exchange for cloth. This exchange might even take place notwithstanding that the commodity imported by Portugal could be produced there with less labour than in England. Though she could make the cloth with the labour of 90 men, she would import it from a country where it required the labour of 100 men to produce it, because it would be advantageous to her rather to employ her capital in the production of wine, for which she would obtain more cloth from England, than she could produce by diverting a portion of her capital from the cultivation of vines to the manufacture of cloth.

Thus England would give the produce of the labour of 100 men for the produce of the labour of 80. Such an exchange could not take place between the individuals of the same country. The labour of 100 Englishmen cannot be given for that of 80 Englishmen, but the produce of the labour of 100 Englishmen may be given for the produce of the labour of 80

Portuguese, 60 Russians, or 120 East Indians. The difference in this respect, between a single country and many, is easily accounted for, by considering the difficulty with which capital moves from one country to another, to seek a more profitable employment, and the activity with which it invariably passes from one province to another in the same country.[1]

It would undoubtedly be advantageous to the capitalists of England, and to the consumers in both countries, that under such circumstances the wine and the cloth should both be made in Portugal, and therefore that the capital and labour of England employed in making cloth should be removed to Portugal for that purpose. In that case, the relative value of these commodities would be regulated by the same principle as if one were the produce of Yorkshire and the other of London: and in every other case, if capital freely flowed towards those countries where it could be most profitably employed, there could be no difference in the rate of profit, and no other difference in the real or labour price of commodities than the additional quantity of labour required to convey them to the various markets where they were to be sold.

Experience, however, shows that the fancied or real insecurity of capital, when not under the immediate control of its owner, together with the natural disinclination which every man has to quit the country of his birth and connections, and intrust himself, with all his habits fixed, to a strange government and new laws, check the emigration of capital. These feelings, which I should be sorry to see weakened, induce most men of property to be satisfied with a low rate of profits in their own country, rather than seek a more advantageous employment for their wealth in foreign nations.

Gold and silver having been chosen for the general medium of circulation, they are, by the competition of commerce, distributed in such proportions amongst the different countries of the world as to accommodate themselves to the natural traffic

[1] It will appear, then, that a country possessing very considerable advantages in machinery and skill, and which may therefore be enabled to manufacture commodities with much less labour than her neighbours, may, in return for such commodities, import a portion of the corn required for its consumption, even if its land were more fertile and corn could be grown with less labour than in the country from which it was imported. Two men can both make shoes and hats, and one is superior to the other in both employments; but in making hats he can only exceed his competitor by one-fifth or 20 per cent., and in making shoes he can excel him by one-third or 33 per cent.;—will it not be for the interest of both that the superior man should employ himself exclusively in making shoes, and the inferior man in making hats?

which would take place if no such metals existed, and the trade between countries were purely a trade of barter.

Thus, cloth cannot be imported into Portugal unless it sell there for more gold than it cost in the country from which it was imported; and wine cannot be imported into England unless it will sell for more there than it cost in Portugal. If the trade were purely a trade of barter, it could only continue whilst England could make cloth so cheap as to obtain a greater quantity of wine with a given quantity of labour by manufacturing cloth than by growing vines; and also whilst the industry of Portugal were attended by the reverse effects. Now suppose England to discover a process for making wine, so that it should become her interest rather to grow it than import it; she would naturally divert a portion of her capital from the foreign trade to the home trade; she would cease to manufacture cloth for exportation, and would grow wine for herself. The money price of these commodities would be regulated accordingly; wine would fall here while cloth continued at its former price, and in Portugal no alteration would take place in the price of either commodity. Cloth would continue for some time to be exported from this country, because its price would continue to be higher in Portugal than here; but money instead of wine would be given in exchange for it, till the accumulation of money here, and its diminution abroad, should so operate on the relative value of cloth in the two countries that it would cease to be profitable to export it. If the improvement in making wine were of a very important description, it might become profitable for the two countries to exchange employments; for England to make all the wine, and Portugal all the cloth consumed by them; but this could be effected only by a new distribution of the precious metals, which should raise the price of cloth in England and lower it in Portugal. The relative price of wine would fall in England in consequence of the real advantage from the improvement of its manufacture; that is to say, its natural price would fall; the relative price of cloth would rise there from the accumulation of money.

Thus, suppose before the improvement in making wine in England the price of wine here were £50 per pipe, and the price of a certain quantity of cloth were £45, whilst in Portugal the price of the same quantity of wine was £45, and that of the same quantity of cloth £50; wine would be exported from Portugal with a profit of £5, and cloth from England with a profit of the same amount.

Suppose that, after the improvement, wine falls to £45 in England, the cloth continuing at the same price. Every transaction in commerce is an independent transaction. Whilst a merchant can buy cloth in England for £45, and sell it with the usual profit in Portugal, he will continue to export it from England. His business is simply to purchase English cloth, and to pay for it by a bill of exchange, which he purchases with Portuguese money. It is to him of no importance what becomes of this money: he has discharged his debt by the remittance of the bill. His transaction is undoubtedly regulated by the terms on which he can obtain this bill, but they are known to him at the time; and the causes which may influence the market price of bills, or the rate of exchange, is no consideration of his.

If the markets be favourable for the exportation of wine from Portugal to England, the exporter of the wine will be a seller of a bill, which will be purchased either by the importer of the cloth, or by the person who sold him his bill; and thus, without the necessity of money passing from either country, the exporters in each country will be paid for their goods. Without having any direct transaction with each other, the money paid in Portugal by the importer of cloth will be paid to the Portuguese exporter of wine; and in England by the negotiation of the same bill the exporter of the cloth will be authorised to receive its value from the importer of wine.

But if the prices of wine were such that no wine could be exported to England, the importer of cloth would equally purchase a bill; but the price of that bill would be higher, from the knowledge which the seller of it would possess that there was no counter bill in the market by which he could ultimately settle the transactions between the two countries; he might know that the gold or silver money which he received in exchange for his bill must be actually exported to his correspondent in England, to enable him to pay the demand which he had authorised to be made upon him, and he might therefore charge in the price of his bill all the expenses to be incurred, together with his fair and usual profit.

If then this premium for a bill on England should be equal to the profit on importing cloth, the importation would of course cease; but if the premium on the bill were only 2 per cent., if to be enabled to pay a debt in England of £100, £102 should be paid in Portugal, whilst cloth which cost £45 would sell for £50, cloth would be imported, bills would be bought, and money would be exported, till the diminution of money in Portugal, and its accumulation in England, had produced such a state of prices as would make it no longer profitable to continue these transactions.

But the diminution of money in one country, and its increase in another, do not operate on the price of one commodity only, but on the prices of all, and therefore the price of wine and cloth will be both raised in England and both lowered in Portugal. The price of cloth, from being £45 in one country and £50 in the other, would probably fall to £49 or £48 in Portugal, and rise to £46 or £47 in England, and not afford a sufficient profit after paying a premium for a bill to induce any merchant to import that commodity.

It is thus that the money of each country is apportioned to it in such quantities only as may be necessary to regulate a profitable trade of barter. England exported cloth in exchange for wine because, by so doing, her industry was rendered more productive to her; she had more cloth and wine than if she had manufactured both for herself; and Portugal imported cloth and exported wine because the industry of Portugal could be more beneficially employed for both countries in producing wine. Let there be more difficulty in England in producing cloth, or in Portugal in producing wine, or let there be more facility in England in producing wine, or in Portugal in producing cloth, and the trade must immediately cease.

No change whatever takes place in the circumstances of Portugal; but England finds that she can employ her labour more productively in the manufacture of wine, and instantly the trade of barter between the two countries changes. Not only is the exportation of wine from Portugal stopped, but a new distribution of the precious metals takes place, and her importation of cloth is also prevented.

Both countries would probably find it their interest to make their own wine and their own cloth; but this singular result would take place: in England, though wine would be cheaper, cloth would be elevated in price, more would be paid for it by the consumer; while in Portugal the consumers, both of cloth and of wine, would be able to purchase those commodities cheaper. In the country where the improvement was made prices would be enhanced; in that where no change had taken place, but where they had been deprived of a profitable branch of foreign trade, prices would fall.

This, however, is only a seeming advantage to Portugal, for the quantity of cloth and wine together produced in that country would be diminished, while the quantity produced in England would be increased. Money would in some degree have changed its value in the two countries; it would be lowered in England and raised in Portugal. Estimated in money, the whole revenue of Portugal would be diminished; estimated in the same medium the whole revenue of England would be increased.

Thus, then, it appears that the improvement of a manufacture in any country tends to alter the distribution of the precious metals amongst the nations of the world: it tends to increase the quantity of commodities, at the same time that it raises general prices in the country where the improvement takes place.

To simplify the question, I have been supposing the trade between two countries to be confined to two commodities—to wine and cloth; but it is well known that many and various articles enter into the list of exports and imports. By the abstraction of money from one country, and the accumulation of it in another, all commodities are affected in price, and consequently encouragement is given to the exportation of many more commodities besides money, which will therefore prevent so great an effect from taking place on the value of money in the two countries as might otherwise be expected.

Beside the improvements in arts and machinery, there are various other causes which are constantly operating on the natural course of trade, and which interfere with the equilibrium and the relative value of money. Bounties on exportation or importation, new taxes on commodities, sometimes by their direct, and at other times by their indirect operation, disturb the natural trade of barter, and produce a consequent necessity of importing or exporting money, in order that prices may be accommodated to the natural course of commerce; and this effect is produced not only in the country where the disturbing cause takes place, but, in a greater or less degree, in every country of the commercial world.

This will in some measure account for the different value of money in different countries; it will explain to us why the prices of home commodities, and those of great bulk, though of comparatively small value, are, independently of other causes, higher in those countries where manufactures flourish. Of two countries having precisely the same population, and the same quantity of land of equal fertility in cultivation, with the same knowledge too of agriculture, the prices of raw produce will be highest in that where the greater skill and the better machinery is used in the manufacture of exportable commodities. The rate of profits will probably differ but little; for wages, or the real reward of the labourer, may be the same in both; but those wages, as well as raw produce, will be rated higher in money in that country, into which, from the advantages attending their skill and machinery, an abundance of money is imported in exchange for their goods.

Of these two countries, if one had the advantage in the manufacture of goods of one quality, and the other in the manufacture of goods of another quality, there would be no decided influx of the precious metals into either; but if the advantage very heavily preponderated in favour of either, that effect would be inevitable.

In the former part of this work, we have assumed, for the purpose of argument, that money always continued of the same value; we are now endeavouring to show that, besides the ordinary variations in the value of money, and those which are common to the whole commercial world, there are also partial variations to which money is subject in particular countries; and to the fact that the value of money is never the same in any two countries, depending as it does on relative taxation, on manufacturing skill, on the advantages of climate, natural productions, and many other causes.

Although, however, money is subject to such perpetual variations, and consequently the prices of the commodities which are common to most countries are also subject to considerable difference, yet no effect will be produced on the rate of profits, either from the influx or efflux of money. Capital will not be increased because the circulating medium is augmented. If the rent paid by the farmer to his landlord, and the wages to his labourers, be 20 per cent. higher in one country than another, and if at the same time the nominal value of the farmer's capital be 20 per cent. more, he will receive precisely the same rate of profits, although he should sell his raw produce 20 per cent. higher.

Profits, it cannot be too often repeated, depend on wages; not on nominal, but real wages; not on the number of pounds that may be annually paid to the labourer, but on the number of days' work necessary to obtain those pounds. Wages may therefore be precisely the same in two countries; they may bear, too, the same proportion to rent, and to the whole produce obtained from the land, although in one of those countries the labourer should receive ten shillings per week and in the other twelve.

In the early states of society, when manufactures have made little progress, and the produce of all countries is nearly similar, consisting of the bulky and most useful commodities, the value of money in different countries will be chiefly regulated by their distance from the mines which supply the precious metals; but as the arts and improvements of society advance, and different nations excel in particular manufactures, although distance will still enter into the calculation, the value of the precious metals will be chiefly regulated by the superiority of those manufactures.

Suppose all nations to produce corn, cattle, and coarse clothing only, and that it was by the exportation of such commodities that gold could be obtained from the countries which produced them, or from those who held them in subjection; gold would naturally be of greater exchangeable value in Poland than in England, on account of the greater expense of sending such a bulky commodity as corn the more distant voyage, and also the greater expense attending the conveying of gold to Poland.

This difference in the value of gold, or, which is the same thing, this difference in the price of corn in the two countries, would exist, although the facilities of producing corn in England should far exceed those of Poland, from the greater fertility of the land and the superiority in the skill and implements of the labourer.

If, however, Poland should be the first to improve her manufactures, if she should succeed in making a commodity which was generally desirable, including great value in little bulk, or if she should be exclusively blessed with some natural production, generally desirable, and not possessed by other countries, she would obtain an additional quantity of gold in exchange for this commodity, which would operate on the price of her corn, cattle, and coarse clothing. The disadvantage of distance would probably be more than compensated by the advantage of having an exportable commodity of great value, and money would be permanently of lower value in Poland than in England. If, on the contrary, the advantage of skill and machinery were possessed by England, another reason would be added to that which before existed why gold should be less valuable in England than in Poland, and why corn, cattle, and clothing should be at a higher price in the former country.

These I believe to be the only two causes which regulate the comparative value of money in the different countries of the world; for although taxation occasions a disturbance of the equilibrium of money, it does so by depriving the country in which it is imposed of some of the advantages attending skill, industry, and climate.

It has been my endeavour carefully to distinguish between a low value of money and a high value of corn, or any other commodity with which money may be compared. These have been generally considered as meaning the same thing; but it is evident that when corn rises from five to ten shillings a bushel, it may be owing either to a fall in the value of money or to a rise in the value of corn. Thus we have seen that, from the necessity of having recourse successively to land of a worse and worse quality, in order to feed an increasing population, corn must rise in relative value to other things. If therefore money continue permanently of the same value, corn will exchange for more of such money, that is to say, it will rise in price. The same rise in the price of corn will be produced by such improvement of machinery in manufactures as shall enable us to manufacture commodities with peculiar advantages: for the influx of money will be the consequence; it will fall in value, and therefore exchange for less corn. But the effects resulting from a high price of corn when produced by the rise in the value of corn, and when caused by a fall in the value of money, are totally different. In both cases the money price of wages will rise, but if it be in consequence of the fall in the value of money, not only wages and corn, but all other commodities will rise. If the manufacturer has more to pay for wages he will receive more for his manufactured goods, and the rate of profits will remain unaffected. But when the rise in the price of corn is the effect of the difficulty of production, profits will fall; for the manufacturer will be obliged to pay more wages, and will not be enabled to remunerate himself by raising the price of his manufactured commodity.

Any improvement in the facility of working the mines, by which the precious metals may be produced with a less quantity of labour, will sink the value of money generally. It will then exchange for fewer commodities in all countries; but when any particular country excels in manufactures, so as to occasion an influx of money towards it, the value of money will be lower, and the prices of corn and labour will be relatively higher in that country than in any other.

This higher value of money will not be indicated by the exchange; bills may continue to be negotiated at par, although the prices of corn and labour should be 10, 20, or 30 per cent.

higher in one country than another. Under the circumstances supposed, such a difference of prices is the natural order of things, and the exchange can only be at par when a sufficient quantity of money is introduced into the country excelling in manufactures, so as to raise the price of its corn and labour. If foreign countries should prohibit the exportation of money, and could successfully enforce obedience to such a law, they might indeed prevent the rise in the prices of the corn and labour of the manufacturing country; for such rise can only take place after the influx of the precious metals, supposing paper money not to be used; but they could not prevent the exchange from being very unfavourable to them. If England were the manufacturing country, and it were possible to prevent the importation of money, the exchange with France, Holland, and Spain might be 5, 10, or 20 per cent. against those countries.

Whenever the current of money is forcibly stopped, and when money is prevented from settling at its just level, there are no limits to the possible variations of the exchange. The effects are similar to those which follow when a paper money, not exchangeable for specie at the will of the holder, is forced into circulation. Such a currency is necessarily confined to the country where it is issued: it cannot, when too abundant, diffuse itself generally amongst other countries. The level of circulation is destroyed, and the exchange will inevitably be unfavourable to the country where it is excessive in quantity: just so would be the effects of a metallic circulation if by forcible means, by laws which could not be evaded, money should be detained in a country, when the stream of trade gave it an impetus towards other countries.

When each country has precisely the quantity of money which it ought to have, money will not indeed be of the same value in each, for with respect to many commodities it may differ 5, 10, or even 20 per cent., but the exchange will be at par. One hundred pounds in England, or the silver which is in £100, will purchase a bill of £100, or an equal quantity of silver in France, Spain, or Holland.

In speaking of the exchange and the comparative value of money in different countries, we must not in the least refer to the value of money estimated in commodities in either country. The exchange is never ascertained by estimating the comparative value of money in corn, cloth, or any commodity whatever, but by estimating the value of the currency of one country in the currency of another.

It may also be ascertained by comparing it with some standard common to both countries. If a bill on England for £100 will purchase the same quantity of goods in France or Spain that a bill on Hamburgh for the same sum will do, the exchange between Hamburgh and England is at par; but if a bill on England for £130 will purchase no more than a bill on Hamburgh for £100, the exchange is 30 per cent. against England.

In England £100 may purchase a bill, or the right of receiving £101 in Holland, £102 in France, and £105 in Spain. The exchange with England is, in that case, said to be 1 per cent. against Holland, 2 per cent. against France, and 5 per cent. against Spain. It indicates that the level of currency is higher than it should be in those countries, and the comparative value of their currencies, and that of England, would be immediately restored to par by extracting from theirs or by adding to that of England.

Those who maintain that our currency was depreciated during the last ten years, when the exchange varied from 20 to 30 per cent. against this country, have never contended, as they have been accused of doing, that money could not be more valuable in one country than another as compared with various commodities; but they did contend that £130 could not be detained in England unless it was depreciated, when it was of no more value, estimated in the money of Hamburgh or of Holland, than the bullion in £100.

By sending 130 good English pounds sterling to Hamburgh, even at an expense of £5, I should be possessed there of £125; what then could make me consent to give £130 for a bill which would give me £100 in Hamburgh, but that my pounds were not good pounds sterling? — they were deteriorated, were degraded in intrinsic value below the pounds sterling of Hamburgh, and if actually sent there, at an expense of £5, would sell only for £100. With metallic pounds sterling, it is not denied that my £130 would procure me £125 in Hamburgh, but with paper pounds sterling I can only obtain £100; and yet it was maintained that £130 in paper was of equal value with £130 in silver or gold.

Some indeed more reasonably maintained that £130 in paper was not of equal value with £130 in metallic money; but they said that it was the metallic money which had changed its value and not the paper money. They wished to confine the meaning of the word depreciation to an actual fall of value, and not to a comparative difference between the value of money and the standard by which by law it is regulated. One hundred pounds of English money was formerly of equal value with and could purchase £100 of Hamburgh money: in any other country a bill of £100 on England, or on Hamburgh, could purchase precisely the same quantity of commodities. To obtain the same things, I was lately obliged to give £130 English money, when Hamburgh could obtain them for £100 Hamburgh money. If English money was of the same value then as before, Hamburgh money must have risen in value. But where is the proof of this? How is it to be ascertained whether English money has fallen or Hamburgh money has risen? there is no standard by which this can be determined. It is a plea which admits of no proof, and can neither be positively affirmed nor positively contradicted. The nations of the world must have been early convinced that there was no standard of value in nature to which they might unerringly refer, and therefore chose a medium which on the whole appeared to them less variable than any other commodity.

To this standard we must conform till the law is changed, and till some other commodity is discovered by the use of which we shall obtain a more perfect standard than that which we have established. While gold is exclusively the standard in this country money will be depreciated when a pound sterling is not of equal value with 5 dwts. and 3 grs. of standard gold, and that whether gold rises or falls in general value.

CHAPTER VIII

ON TAXES

TAXES are a portion of the produce of the land and labour of a country placed at the disposal of the government; and are always ultimately paid either from the capital or from the revenue of the country.

We have already shown how the capital of a country is either fixed or circulating, according as it is of a more or of a less durable nature. It is difficult to define strictly where the distinction between circulating and fixed capital begins; for there are almost infinite degrees in the durability of capital. The food of a country is consumed and reproduced at least once in every year, the clothing of the labourer is probably not consumed and reproduced in less than two years; whilst his house and furniture are calculated to endure for a period of ten or twenty years.

When the annual productions of a country more than replace its annual consumption, it is said to increase its capital; when its annual consumption is not at least replaced by its annual production, it is said to diminish its capital. Capital may therefore be increased by an increased production, or by a diminished unproductive consumption.

If the consumption of the government when increased by the levy of additional taxes be met either by an increased production or by a diminished consumption on the part of the people, the taxes will fall upon revenue, and the national capital will remain unimpaired; but if there be no increased production or diminished unproductive consumption on the part of the people, the taxes will necessarily fall on capital, that is to say, they will impair the fund allotted to productive consumption.[1]

[1] It must be understood that all the productions of a country are consumed; but it makes the greatest difference imaginable whether they are consumed by those who reproduce or by those who do not reproduce another value. When we say that revenue is saved and added to capital, what we mean is, that the portion of revenue, so said to be added to capital, is consumed by productive instead of unproductive labourers. There can be no greater error than in supposing that capital is increased by non-consumption. If the price of labour should rise so high that, notwith-

In proportion as the capital of a country is diminished, its productions will be necessarily diminished; and, therefore, if the same unproductive expenditure on the part of the people and of the government continue, with a constantly diminishing annual reproduction, the resources of the people and the state will fall away with increasing rapidity, and distress and ruin will follow.

Notwithstanding the immense expenditure of the English government during the last twenty years, there can be little doubt but that the increased production on the part of the people has more than compensated for it. The national capital has not merely been unimpaired, it has been greatly increased, and the annual revenue of the people, even after the payment of their taxes, is probably greater at the present time than at any former period of our history.

For the proof of this, we might refer to the increase of population — to the extension of agriculture — to the increase of shipping and manufactures—to the building of docks—to the opening of numerous canals, as well as to many other expensive undertakings; all denoting an increase both of capital and of annual production.

Still, however, it is certain that, but for taxation, this increase of capital would have been much greater. There are no taxes which have not a tendency to lessen the power to accumulate. All taxes must either fall on capital or revenue. If they encroach on capital, they must proportionably diminish that fund by whose extent the extent of the productive industry of the country must always be regulated; and if they fall on revenue, they must either lessen accumulation, or force the contributors to save the amount of the tax, by making a corresponding diminution of their former unproductive consumption of the necessaries and luxuries of life. Some taxes will produce these effects in a much greater degree than others; but the great evil of taxation is to be found, not so much in any selection of its objects, as in the general amount of its effects taken collectively.

Taxes are not necessarily taxes on capital because they are laid on capital; nor on income because they are laid on income. If from my income of £1000 per annum I am required to pay £100, it will really be a tax on my income should I be content with the expenditure of the remaining £900; but it will be a tax on capital if I continue to spend £1000.

standing the increase of capital, no more could be employed, I should say that such increase of capital would be still unproductively consumed.

The capital from which my income of £1000 is derived may be of the value of £10,000; a tax of one per cent. on such capital would be £100; but my capital would be unaffected if, after paying this tax, I in like manner contented myself with the expenditure of £900.

The desire which every man has to keep his station in life, and to maintain his wealth at the height which it has once attained, occasions most taxes, whether laid on capital or on income, to be paid from income; and, therefore, as taxation proceeds, or as government increases its expenditure, the annual enjoyments of the people must be diminished, unless they are enabled proportionally to increase their capitals and income. It should be the policy of governments to encourage a disposition to do this in the people, and never to lay such taxes as will inevitably fall on capital; since, by so doing, they impair the funds for the maintenance of labour, and thereby diminish the future production of the country.

In England this policy has been neglected in taxing the probates of wills, in the legacy duty, and in all taxes affecting the transference of property from the dead to the living. If a legacy of £1000 be subject to a tax of £100, the legatee considers his legacy as only £900 and feels no particular motive to save the £100 duty from his expenditure, and thus the capital of the country is diminished; but if he had really received £1000, and had been required to pay £100 as a tax on income, on wine, on horses, or on servants, he would probably have diminished, or rather not increased his expenditure by that sum, and the capital of the country would have been unimpaired.

" Taxes upon the transference of property from the dead to the living," says Adam Smith, " fall finally, as well as immediately, upon the persons to whom the property is transferred. Taxes on the sale of land fall altogether upon the seller. The seller is almost always under the necessity of selling, and must, therefore, take such a price as he can get. The buyer is scarce ever under the necessity of buying, and will, therefore, only give such a price as he likes. He considers what the land will cost him in tax and price together. The more he is obliged to pay in the way of tax, the less he will be disposed to give in the way of price. Such taxes, therefore, fall almost always upon a necessitous person, and must, therefore, be very cruel and oppressive." " Stamp duties, and duties upon the registration of bonds and contracts for borrowed money, fall altogether upon the borrower, and in fact are always paid by him. Duties of the same kind upon law proceedings fall upon the suitors. They reduce to both the capital value of the subject in dispute. The more it costs to acquire any property, the less must be the net value of it when acquired. All taxes upon the transference of property of every kind, so far as they diminish the capital value of that property, tend to diminish the funds destined for the maintenance of labour. They are all more or less unthrifty taxes that increase the revenue of the sovereign, which seldom maintains any but unproductive labourers, at the expense of the capital of the people, which maintains none but productive."

But this is not the only objection to taxes on the transference of property; they prevent the national capital from being distributed in the way most beneficial to the community. For the general prosperity there cannot be too much facility given to the conveyance and exchange of all kinds of property, as it is by such means that capital of every species is likely to find its way into the hands of those who will best employ it in increasing the productions of the country. "Why," asks M. Say, " does an individual wish to sell his land? it is because he has another employment in view in which his funds will be more productive. Why does another wish to purchase this same land? it is to employ a capital which brings him in too little, which was unemployed, or the use of which he thinks susceptible of improvement. This exchange will increase the general income, since it increases the income of these parties. But if the charges are so exorbitant as to prevent the exchange, they are an obstacle to this increase of the general income." Those taxes, however, are easily collected; and this by many may be thought to afford some compensation for their injurious effects.

CHAPTER IX

HAVING in a former part of this work established, I hope satisfactorily, the principle that the price of corn is regulated by the cost of its production on that land exclusively, or rather with that capital exclusively, which pays no rent, it will follow that whatever may increase the cost of production will increase the price; whatever may reduce it will lower the price. The necessity of cultivating poorer land, or of obtaining a less return with a given additional capital on land already in cultivation, will inevitably raise the exchangeable value of raw produce. The discovery of machinery, which will enable the cultivator to obtain his corn at a less cost of production, will necessarily lower its exchangeable value. Any tax which may be imposed on the cultivator, whether in the shape of land-tax, tithes, or a tax on the produce when obtained, will increase the cost of production, and will therefore raise the price of raw produce.

If the price of raw produce did not rise so as to compensate the cultivator for the tax, he would naturally quit a trade where his profits were reduced below the general level of profits; this would occasion a diminution of supply, until the unabated demand should have produced such a rise in the price of raw produce as to make the cultivation of it equally profitable with the investment of capital in any other trade.

A rise of price is the only means by which he could pay the tax, and continue to derive the usual and general profits from this employment of his capital. He could not deduct the tax from his rent, and oblige his landlord to pay it, for he pays no rent. He would not deduct it from his profits, for there is no reason why he should continue in an employment which yields small profits, when all other employments are yielding greater. There can then be no question but that he will have the power of raising the price of raw produce by a sum equal to the tax.

A tax on raw produce would not be paid by the landlord; it would not be paid by the farmer; but it would be paid, in an increased price, by the consumer.

Rent, it should be remembered, is the difference between the produce obtained by equal portions of labour and capital employed on land of the same or different qualities. It should be remembered, too, that the money rent of land, and the corn rent of land, do not vary in the same proportion.

In the case of a tax on raw produce, of a land-tax, or tithes, the corn rent of land will vary, while the money rent will remain as before.

If, as we have before supposed, the land in cultivation were of three qualities, and that with an equal amount of capital,

180 qrs. of corn were obtained from land No. 1

170 . . . from . . . 2

160 . . . from . . . 3

the rent of No. 1 would be 20 quarters, the difference between that of No. 3 and No. 1; and of No. 2, 10 quarters, the difference between that of No. 3 and No. 2; while No. 3 would pay no rent whatever.

Now, if the price of corn were £4 per quarter, the money rent of No. 1 would be £80, and that of No. 2, £40.

Suppose a tax of 8s. per quarter to be imposed on corn; then the price would rise to £4 8s.; and if the landlords obtained the same corn rent as before, the rent of No. 1 would be £88 and that of No. 2, £44. But they would not obtain the same corn rent; the tax would fall heavier on No. 1 than on No. 2, and on No. 2 than on No. 3, because it would be levied on a greater quantity of corn. It is the difficulty of production on No. 3 which regulates price; and corn rises to £4 8s., that the profits of the capital employed on No. 3 may be on a level with the general profits of stock.

The produce and tax on the three qualities of land will be as follows:

No. 1, yielding	180	qrs. at £4 8s. per qr. .	£792
Deduct the value of	16.3	or 8s. per qr. on 180 qrs. .	72
Net corn produce	163.7	Net money produce	£720

No. 2, yielding	170	qrs. at £4 8s. per qr. .	£748
Deduct the value of	15.4	{ qrs. at £4 8s. or 8s. per qr. on 170 qrs. }	68
Net corn produce	154.6	Net money produce	£680

No. 3, yielding	160	qrs. at £4 8s. . . .	£704
Deduct the value of	14.5	{ qrs. at £4 8s. or 8s. per qr. on 160 }	64
Net corn produce	145.5	Net money produce	£640

The money rent of No. 1 would continue to be £80, or the difference between £640 and £720; and that of No. 2, £40, or the difference between £640 and £680, precisely the same as before; but the corn rent will be reduced from 20 quarters on No. 1, to 18.2 quarters, the difference between 145.5 and 163.7 quarters, and that on No. 2 from 10 to 9.1 quarters, the difference between 145.5 and 154.6 quarters.

A tax on corn, then, would fall on the consumers of corn, and would raise its value, as compared with all other commodities, in a degree proportioned to the tax. In proportion as raw produce entered into the composition of other commodities would their value also be raised, unless the tax were countervailed by other causes. They would in fact be indirectly taxed, and their value would rise in proportion to the tax.

A tax, however, on raw produce, and on the necessaries of the labourer, would have another effect — it would raise wages. From the effect of the principle of population on the increase of mankind, wages of the lowest kind never continue much above that rate which nature and habit demand for the support of the labourers. This class is never able to bear any considerable proportion of taxation; and, consequently, if they had to pay 8s. per quarter in addition for wheat, and in some smaller proportion for other necessaries, they would not be able to subsist on the same wages as before, and to keep up the race of labourers. Wages would inevitably and necessarily rise; and, in proportion as they rose, profits would fall. Government would receive a tax of 8s. per quarter on all the corn consumed in the country, a part of which would be paid directly by the consumers of corn; the other part would be paid indirectly by those who employed labour, and would affect profits in the same manner as if wages had been raised from the increased demand for labour compared with the supply, or from an increasing difficulty of obtaining the food and necessaries required by the labourer.

In as far as the tax might affect consumers it would be an equal tax, but in as far as it would affect profits it would be a partial tax; for it would neither operate on the landlord nor on the stockholder, since they would continue to receive, the one the same money rent, the other the same money dividends as before. A tax on the produce of the land then would operate as follows:—

1st, It would raise the price of raw produce by a sum equal to the tax, and would therefore fall on each consumer in proportion to his consumption.

2nd, It would raise the wages of labour, and lower profits.

It may then be objected against such a tax,

1st, That by raising the wages of labour, and lowering profits, it is an unequal tax, as affects the income of the farmer, trader, and manufacturer, and leaves untaxed the income of the landlord, stockholder, and others enjoying fixed incomes.

2nd, That there would be a considerable interval between the rise in the price of corn and the rise of wages, during which much distress would be experienced by the labourer.

3rd, That raising wages and lowering profits is a discouragement to accumulation, and acts in the same way as a natural poverty of soil.

4th, That by raising the price of raw produce, the prices of all commodities into which raw produce enters would be raised, and that therefore we should not meet the foreign manufacturer on equal terms in the general market.

With respect to the first objection, that by raising the wages of labour and lowering profits, it acts unequally, as it affects the income of the farmer, trader, and manufacturer, and leaves untaxed the income of the landlord, stockholder, and others enjoying fixed incomes—it may be answered that if the operation of the tax be unequal it is for the legislature to make it equal, by taxing directly the rent of land and the dividends from stock. By so doing, all the objects of an income tax would be obtained without the inconvenience of having recourse to the obnoxious measure of prying into every man's concerns, and arming commissioners with powers repugnant to the habits and feelings of a free country.

With respect to the second objection, that there would be a considerable interval between the rise of the price of corn and the rise of wages, during which much distress would be experienced by the lower classes—I answer that under different circumstances, wages follow the price of raw produce with very different degrees of celerity; that in some cases no effect whatever is produced on wages by a rise of corn; in others, the rise of wages precedes the rise in the price of corn; again, in some the effect on wages is slow, and in others rapid.

Those who maintain that it is the price of necessaries which regulates the price of labour, always allowing for the particular state of progression in which the society may be, seem to have conceded too readily that a rise or fall in the price of necessaries will be very slowly succeeded by a rise or fall of wages. A high price of provisions may arise from very different causes, and may accordingly produce very different effects. It may arise from

1st, A deficient supply.
2nd, From a gradually increasing demand, which may be ultimately attended with an increased cost of production.
3rd, From a fall in the value of money.
4th, From taxes on necessaries.

These four causes have not been sufficiently distinguished and separated by those who have inquired into the influence of a high price of necessaries on wages. We will examine them severally.

A bad harvest will produce a high price of provisions, and the high price is the only means by which the consumption is compelled to conform to the state of the supply. If all the purchasers of corn were rich, the price might rise to any degree, but the result would remain unaltered; the price would at last be so high, that the least rich would be obliged to forego the use of a part of the quantity which they usually consumed, as by diminished consumption alone the demand could be brought down to the limits of the supply. Under such circumstances no policy can be more absurd than that of forcibly regulating money wages by the price of food, as is frequently done, by misapplication of the poor laws. Such a measure affords no real relief to the labourer, because its effect is to raise still higher the price of corn, and at last he must be obliged to limit his consumption in proportion to the limited supply. In the natural course of affairs a deficient supply from bad seasons, without any pernicious and unwise interference, would not be followed by a rise of wages. The raising of wages is merely nominal to those who receive them; it increases the competition in the corn market, and its ultimate effect is to raise the profits of the growers and dealers in corn. The wages of labour are really regulated by the proportion between the supply and demand of necessaries, and the supply and demand of labour; and money is merely the medium, or measure, in which wages are expressed. In this case, then, the distress of the labourer is unavoidable, and no legislation can afford a remedy, except by the importation of additional food or by adopting the most useful substitutes.

When a high price of corn is the effect of an increasing demand, it is always preceded by an increase of wages, for demand cannot increase without an increase of means in the people to pay for that which they desire. An accumulation of capital naturally produces an increased competition among the employers of labour, and a consequent rise in its price. The increased wages are not always immediately expended on food, but are first made to contribute to the other enjoyments of the labourer. His improved condition, however, induces and enables him to marry, and then the demand for food for the support of his family naturally supersedes that of those other enjoyments on which his wages were temporarily expended. Corn rises, then, because the demand for it increases, because there are those in the society who have improved means of paying for it; and the profits of the farmer will be raised above the general level of profits, till the requisite quantity of capital has been employed on its production. Whether, after this has taken place, corn shall again fall to its former price, or shall continue permanently higher, will depend on the quality of the land from which the increased quantity of corn has been supplied. If it be obtained from land of the same fertility as that which was last in cultivation, and with no greater cost of labour, the price will fall to its former state; if from poorer land, it will continue permanently higher. The high wages in the first instance proceeded from an increase in the demand for labour: inasmuch as it encouraged marriage, and supported children, it produced the effect of increasing the supply of labour. But when the supply is obtained, wages will again fall to their former price, if corn has fallen to its former price: to a higher than the former price, if the increased supply of corn has been produced from land of an inferior quality. A high price is by no means incompatible with an abundant supply: the price is permanently high, not because the quantity is deficient, but because there has been an increased cost in producing it. It generally happens, indeed, that when a stimulus has been given to population, an effect is produced beyond what the case requires; the population may be, and generally is, so much increased as, notwithstanding the increased demand for labour, to bear a greater proportion to the funds for maintaining labourers than before the increase of capital. In this case a reaction will take place, wages will be below their natural level, and will continue so, till the usual proportion between the supply and demand has been restored. In this case, then, the rise in the price of corn is preceded by a rise of wages, and therefore entails no distress on the labourer.

A fall in the value of money, in consequence of an influx of the precious metals from the mines, or from the abuse of the privileges of banking, is another cause for the rise of the price of food; but it will make no alteration in the quantity produced. It leaves undisturbed too the number of labourers, as well as the demand for them; for there will be neither an increase nor a diminution of capital. The quantity of necessaries to be allotted to the labourer depends on the comparative demand and supply of necessaries, with the comparative demand and supply of labour; money being only the medium in which the quantity is expressed; and as neither of these is altered, the real reward of the labourer will not alter. Money wages will rise, but they will only enable him to furnish himself with the same quantity of necessaries as before. Those who dispute this principle are bound to show why an increase of money should not have the same effect in raising the price of labour, the quantity of which has not been increased, as they acknowledge it would have on the price of shoes, of hats, and of corn, if the quantity of those commodities were not increased. The relative market value of hats and shoes is regulated by the demand and supply of hats, compared with the demand and supply of shoes, and money is but the medium in which their value is expressed. If shoes be doubled in price, hats will also be doubled in price, and they will retain the same comparative value. So if corn and all the necessaries of the labourer be doubled in price, labour will be doubled in price also; and while there is no interruption to the usual demand and supply of necessaries and of labour, there can be no reason why they should not preserve their relative value.

Neither a fall in the value of money, nor a tax on raw produce, though each will raise the price, will *necessarily* interfere with the quantity of raw produce, or with the number of people, who are both able to purchase and willing to consume it. It is very easy to perceive why, when the capital of a country increases irregularly, wages should rise, whilst the price of corn remains stationary, or rises in a less proportion; and why, when the capital of a country diminishes, wages should fall whilst corn remains stationary, or falls in a much less proportion, and this too for a considerable time; the reason is, because labour is a commodity which cannot be increased and diminished at pleasure. If there are too few hats in the market for the demand the price will rise, but only for a short time; for in the course of one year, by employing more capital in that trade, any reasonable addition may be made to the quantity of hats, and therefore their market price cannot long very much exceed their natural price; but it is not so with men; you cannot increase their number in one or two years when there is an increase of capital, nor can you rapidly diminish their number when capital is in a retrograde state; and, therefore, the number of hands increasing or diminishing slowly, whilst the funds for the maintenance of labour increase or diminish rapidly, there must be a considerable interval before the price of labour is exactly regulated by the price of corn and necessaries; but in the case of a fall in the value of money, or of a tax on corn, there is not necessarily any excess in the supply of labour, nor any abatement of demand, and therefore there can be no reason why the labourer should sustain a real diminution of wages.

A tax on corn does not necessarily diminish the quantity of corn, it only raises its money price; it does not necessarily diminish the demand compared with the supply of labour; why then should it diminish the portion paid to the labourer? Suppose it true that it did diminish the quantity given to the labourer, in other words, that it did not raise his money wages in the same proportion as the tax raised the price of the corn which he consumed; would not the supply of corn exceed the demand? — would it not fall in price? and would not the labourer thus obtain his usual portion? In such case, indeed, capital would be withdrawn from agriculture; for if the price were not increased by the whole amount of the tax, agricultural profits would be lower than the general level of profits, and capital would seek a more advantageous employment. In regard, then, to a tax on raw produce, which is the point under discussion, it appears to me that no interval which could bear oppressively on the labourer would elapse between the rise in the price of raw produce and the rise in the wages of the labourer; and that therefore no other inconvenience would be suffered by this class than that which they would suffer from any other mode of taxation, namely, the risk that the tax might infringe on the funds destined for the maintenance of labour, and might therefore check or abate the demand for it.

With respect to the third objection against taxes on raw

produce, namely, that the raising wages, and lowering profits, is a discouragement to accumulation, and acts in the same way as a natural poverty of soil; I have endeavoured to show in another part of this work that savings may be as effectually made from expenditure as from production; from a reduction in the value of commodities as from a rise in the rate of profits. By increasing my profits from £1000 to £1200, whilst prices continue the same, my power of increasing my capital by savings is increased, but it is not increased so much as it would be if my profits continued as before, whilst commodities were so lowered in price that £800 would procure me as much as £1000 purchased before.

Now the sum required by the tax must be raised, and the question simply is, whether the same amount shall be taken from individuals by diminishing their profits, or by raising the prices of the commodities on which their profits will be expended.

Taxation under every form presents but a choice of evils; if it do not act on profit, or other sources of income, it must act on expenditure; and provided the burthen be equally borne, and do not repress reproduction, it is indifferent on which it is laid. Taxes on production, or on the profits of stock, whether applied immediately to profits or indirectly by taxing the land or its produce, have this advantage over other taxes; that, provided all other income be taxed, no class of the community can escape them, and each contributes according to his means.

From taxes on expenditure a miser may escape; he may have an income of £10,000 per annum, and expend only £300; but from taxes on profits, whether direct or indirect, he cannot escape; he will contribute to them either by giving up a part, or the value of a part, of his produce; or by the advanced prices of the necessaries essential to production he will be unable to continue to accumulate at the same rate. He may, indeed, have an income of the same value, but he will not have the same command of labour, nor of an equal quantity of materials on which such labour can be exercised.

If a country is insulated from all others, having no commerce with any of its neighbours, it can in no way shift any portion of its taxes from itself. A portion of the produce of its land and labour will be devoted to the service of the state; and I cannot but think that, unless it presses unequally on that class which accumulates and saves, it will be of little importance whether the taxes be levied on profits, on agricultural, or on manufactured commodities. If my revenue be £1000 per annum, and I must pay taxes to the amount of £100, it is of little importance whether I pay it from my revenue, leaving myself only £900, or pay £100 in addition for my agricultural commodities, or for my manufactured goods. If £100 is my fair proportion of the expenses of the country, the virtue of taxation consists in making sure that I shall pay that £100, neither more nor less; and that cannot be effected in any manner so securely as by taxes on wages, profits, or raw produce.

The fourth and last objection which remains to be noticed is: That by raising the price of raw produce, the prices of all commodities into which raw produce enters will be raised, and that, therefore, we shall not meet the foreign manufacturer on equal terms in the general market.

In the first place, corn and *all* home commodities could not be materially raised in price without an influx of the precious metals; for the same quantity of money could not circulate the same quantity of commodities at high as at low prices, and the precious metals never could be purchased with dear commodities. When more gold is required, it must be obtained by giving more and not fewer commodities in exchange for it. Neither could the want of money be supplied by paper, for it is not paper that regulates the value of gold as a commodity, but gold that regulates the value of paper. Unless, then, the value of gold could be lowered, no paper could be added to the circulation without being depreciated. And that the value of gold could not be lowered appears clear when we consider that the value of gold as a commodity must be regulated by the quantity of goods which must be given to foreigners in exchange for it. When gold is cheap, commodities are dear; and when gold is dear, commodities are cheap, and fall in price. Now as no cause is shown why foreigners should sell their gold cheaper than usual, it does not appear probable that there would be any influx of gold. Without such an influx there can be no increase of quantity, no fall in its value, no rise in the general price of goods.[1]

The probable effect of a tax on raw produce would be to raise the price of raw produce, and of all commodities in which raw produce entered, but not in any degree proportioned to the tax; while other commodities in which no raw produce entered,

---

[1] It may be doubted whether commodities, raised in price merely by taxation, would require any more money for their circulation. I believe they would not.

such as articles made of the metals and the earths, would fall in price: so that the same quantity of money as before would be adequate to the whole circulation.

A tax which should have the effect of raising the price of all home productions would not discourage exportation, except during a very limited time. If they were raised in price at home, they could not indeed immediately be profitably exported, because they would be subject to a burthen here from which abroad they were free. The tax would produce the same effect as an alteration in the value of money, which was not general and common to all countries, but confined to a single one. If England were that country, she might not be able to sell, but she would be able to buy, because importable commodities would not be raised in price. Under these circumstances nothing but money could be exported in return for foreign commodities, but this is a trade which could not long continue; a nation cannot be exhausted of its money, for after a certain quantity has left it, the value of the remainder will rise, and such a price of commodities will be the consequence that they will again be capable of being profitably exported. When money had risen, therefore, we should no longer export it in return for goods, but we should export those manufactures which had first been raised in price by the rise in the price of the raw produce from which they were made, and then again lowered by the exportation of money.

But it may be objected that when money so rose in value it would rise with respect to foreign as well as home commodities, and therefore that all encouragement to import foreign goods would cease. Thus, suppose we imported goods which cost £100 abroad, and which sold for £120 here, we should cease to import them when the value of money had so risen in England that they would only sell for £100 here: this, however, could never happen. The motive which determines us to import a commodity is the discovery of its relative cheapness abroad: it is the comparison of its price abroad with its price at home. If a country export hats, and import cloth, it does so because it can obtain more cloth by making hats and exchanging them for cloth than if it made the cloth inself. If the rise of raw produce occasions any increased cost of production in making hats, it would occasion also an increased cost in making cloth. If, therefore, both commodities were made at home, they would both rise. One, however, being a commodity which we import, would not rise, neither would it fall when the value of money rose; for by not falling it would regain its natural relation to the exported commodity. The rise of raw produce makes a hat rise from 30s. to 33s., or 10 per cent.: the same cause, if we manufactured cloth, would make it rise from 20s. to 22s. per yard. This rise does not destroy the relation between cloth and hats; a hat was, and continues to be, worth one yard and a half of cloth. But if we import cloth, its price will continue uniformly at 20s. per yard, unaffected first by the fall, and then by the rise in the value of money; whilst hats, which had risen from 30s. to 33s., will again fall from 33s. to 30s., at which point the relation between cloth and hats will be restored.

To simplify the consideration of this subject, I have been supposing that a rise in the value of raw materials would affect, in an equal proportion, all home commodities; that if the effect on one were to raise it 10 per cent., it would raise all 10 per cent.; but as the value of commodities is very differently made up of raw material and labour; as some commodities, for instance, all those made from the metals, would be unaffected by the rise of raw produce from the surface of the earth, it is evident that there would be the greatest variety in the effects produced on the value of commodities by a tax on raw produce. As far as this effect was produced, it would stimulate or retard the exportation of particular commodities, and would undoubtedly be attended with the same inconvenience that attends the taxing of commodities; it would destroy the natural relation between the value of each. Thus the natural price of a hat, instead of being the same as a yard and a half of cloth, might only be of the value of a yard and a quarter, or it might be of the value of a yard and three quarters, and therefore rather a different direction might be given to foreign trade. All these inconveniences would probably not interfere with the value of the exports and imports; they would only prevent the very best distribution of the capital of the whole world, which is never so well regulated as when every commodity is freely allowed to settle at its natural price, unfettered by artificial restraints.

Although, then, the rise in the price of most of our own commodities would for a time check exportation generally, and might permanently prevent the exportation of a few commodities, it could not materially interfere with foreign trade, and would not place us under any comparative disadvantage as far as regarded competition in foreign markets.

CHAPTER X

A TAX on rent would affect rent only; it would fall wholly on landlords, and could not be shifted to any class of consumers. The landlord could not raise his rent, because he would leave unaltered the difference between the produce obtained from the least productive land in cultivation, and that obtained from land of every other quality. Three sorts of land, No. 1, 2, and 3, are in cultivation, and yield respectively, with the same labour, 180, 170, and 160 quarters of wheat; but No. 3 pays no rent, and is therefore untaxed: the rent then of No. 2 cannot be made to exceed the value of ten, nor No. 1 of twenty quarters. Such a tax could not raise the price of raw produce, because, as the cultivator of No. 3 pays neither rent nor tax, he would in no way be enabled to raise the price of the commodity produced. A tax on rent would not discourage the cultivation of fresh land, for such land pays no rent, and would be untaxed. If No. 4 were taken into cultivation, and yielded 150 quarters, no tax would be paid for such land; but it would create a rent of ten quarters on No. 3, which would then commence paying the tax.

A tax on rent, as rent is constituted, would discourage cultivation, because it would be a tax on the profits of the landlord. The term rent of land, as I have elsewhere observed, is applied to the whole amount of the value paid by the farmer to his landlord, a part only of which is strictly rent. The buildings and fixtures, and other expenses paid for by the landlord, form strictly a part of the stock of the farm, and must have been furnished by the tenant, if not provided by the landlord. Rent is the sum paid to the landlord for the use of the land, and for the use of the land only. The further sum that is paid to him under the name of rent is for the use of the buildings, etc., and is really the profits of the landlord's stock. In taxing rent, as no distinction would be made between that part paid for the use of the land, and that paid for the use of the landlord's stock, a portion of the tax would fall on the landlord's profits, and would, therefore, discourage cultivation, unless the price of raw produce rose. On that land, for the use of which no rent was paid, a compensation under that name might be given to the landlord for the use of his buildings. These buildings would not be erected, nor would raw produce be grown on such land, till the price at which it sold would not only pay for all the usual outgoings, but also this additional one of the tax. This part of the tax does not fall on the landlord, nor on the farmer, but on the consumer of raw produce.

There can be little doubt but that if a tax were laid on rent, landlords would soon find a way to discriminate between that which is paid to them for the use of the land, and that which is paid for the use of the buildings, and the improvements which are made by the landlord's stock. The latter would either be called the rent of house and buildings, or on all new land taken into cultivation such buildings would be erected and improvements would be made by the tenant and not by the landlord. The landlord's capital might indeed be really employed for that purpose; it might be nominally expended by the tenant, the landlord furnishing him with the means, either in the shape of a loan, or in the purchase of an annuity for the duration of the lease. Whether distinguished or not, there is a real difference between the nature of the compensations which the landlord receives for these different objects; and it is quite certain that a tax on the real rent of land falls wholly on the landlord, but that a tax on that remuneration which the landlord receives for the use of this stock expended on the farm, falls, in a progressive country, on the consumer of raw produce. If a tax were laid on rent, and no means of separating the remuneration now paid by the tenant to the landlord under the name of rent were adopted, the tax, as far as it regarded the rent on the buildings and other fixtures, would never fall for any length of time on the landlord, but on the consumer. The capital expended on these buildings, etc., must afford the usual profit of stock; but it would cease to afford this profit on the land last cultivated if the expenses of those buildings, etc., did not fall on the tenant; and if they did, the tenant would then cease to make his usual profits of stock, unless he could charge them on the consumer.

CHAPTER XI

TITHES

TITHES are a tax on the gross produce of the land, and, like taxes on raw produce, fall wholly on the consumer. They differ from a tax on rent, inasmuch as they affect land which such a tax would not reach; and raise the price of raw produce which that tax would not alter. Lands of the worst quality, as well as of the best, pay tithes, and exactly in proportion to the quantity of produce obtained from them; tithes are therefore an equal tax.

If land of the last quality, or that which pays no rent, and which regulates the price of corn, yield a sufficient quantity to give the farmer the usual profits of stock, when the price of wheat is £4 per quarter, the price must rise to £4 8s. before the same profits can be obtained after the tithes are imposed, because for every quarter of wheat the cultivator must pay eight shillings to the church, and if he does not obtain the same profits, there is no reason why he should not quit his employment, when he can get them in other trades.

The only difference between tithes and taxes on raw produce is that one is a variable money tax, the other a fixed money tax. In a stationary state of society, where there is neither increased nor diminished facility of producing corn, they will be precisely the same in their effects; for, in such a state, corn will be at an invariable price, and the tax will therefore be also invariable. In either a retrograde state, or in a state in which great improvements are made in agriculture, and where consequently raw produce will fall in value comparatively with other things, tithes will be a lighter tax than a permanent money tax; for if the price of corn should fall from £4 to £3, the tax would fall from eight to six shillings. In a progressive state of society, yet without any marked improvements in agriculture, the price of corn would rise, and tithes would be a heavier tax than a permanent money tax. If corn rose from £4 to £5, the tithes on the same land would advance from eight to ten shillings.

Neither tithes nor a money tax will affect the money rent of landlords, but both will materially affect corn rents. We have already observed how a money tax operates on corn rents, and it is equally evident that a similar effect would be produced by tithes. If the lands, No. 1, 2, 3, respectively produced 180, 170, and 160 quarters, the rents might be on No. 1, twenty quarters, and on No. 2, ten quarters; but they would no longer preserve that proportion after the payment of tithes; for if a tenth be taken from each, the remaining produce will be 162, 153, 144, and consequently the corn rent of No. 1 will be reduced to eighteen, and that of No. 2 to nine quarters. But the price of corn would rise from £4 to £4 8s. 10⅔d.; for 144 quarters are to £4 as 160 quarters to £4 8s. 10⅔d. and consequently the money rent would continue unaltered; for on No. 1 it would be £80,[1] and on No. 2, £40.[2]

The chief objection against tithes is that they are not a permanent and fixed tax, but increase in value in proportion as the difficulty of producing corn increases. If those difficulties should make the price of corn £4, the tax is 8s.; if they should increase it to £5, the tax is 10s.; and at £6 it is 12s. They not only rise in value, but they increase in amount: thus, when No. 1 was cultivated, the tax was only levied on 180 quarters; when No. 2 was cultivated, it was levied on 180 + 170, or 350 quarters; and when No. 3 was cultivated, on 180 + 170 + 160 = 510 quarters. Not only is the amount of tax increased from 100,000 quarters to 200,000 quarters when the produce is increased from one to two millions of quarters; but, owing to the increased labour necessary to produce the second million, the relative value of raw produce is so advanced that the 200,000 quarters may be, though only twice in quantity, yet in value three times that of the 100,000 quarters which were paid before.

If an equal value were raised for the church by any other means, increasing in the same manner as tithes increase, proportionably with the difficulty of cultivation, the effect would be the same; and therefore it is a mistake to suppose that, because they are raised on the land, they discourage cultivation more than an equal amount would do if raised in any other manner. The church would in both cases be constantly obtaining an increased portion of the net produce of the land and labour of the country. In an improving state of society, the net produce of land is always diminishing in proportion to its gross produce; but it is from the net income of a country that

[1] 18 quarters at £4 8s. 10⅔d.　　　　[2] 9 quarters at £4 8s. 10⅔d.

all taxes are ultimately paid, either in a progressive or in a stationary country. A tax increasing with the gross income, and falling on the net income, must necessarily be a very burdensome and a very intolerable tax. Tithes are a tenth of the gross and not of the net produce of the land, and therefore as society improves in wealth, they must, though the same proportion of the gross produce, become a larger and larger proportion of the net produce.

Tithes, however, may be considered as injurious to landlords, inasmuch as they act as a bounty on importation, by taxing the growth of home corn while the importation of foreign corn remains unfettered. And if, in order to relieve the landlords from the effects of the diminished demand for land which such a bounty must encourage, imported corn were also taxed, in an equal degree with corn grown at home, and the produce paid to the state, no measure could be more fair and equitable; since whatever were paid to the state by this tax would go to diminish the other taxes which the expenses of government make necessary; but if such a tax were devoted only to increase the fund paid to the church, it might indeed on the whole increase the general mass of production, but it would diminish the portion of that mass allotted to the productive classes.

If the trade of cloth were left perfectly free, our manufacturers might be able to sell cloth cheaper than we could import it. If a tax were laid on the home manufacturer, and not on the importer of cloth, capital might be injuriously driven from the manufacture of cloth to the manufacture of some other commodity, as cloth might then be imported cheaper than it could be made at home. If imported cloth should also be taxed, cloth would again be manufactured at home. The consumer first bought cloth at home because it was cheaper than foreign cloth; he then bought foreign cloth because it was cheaper untaxed than home cloth taxed: he lastly bought it again at home because it was cheaper when both home and foreign cloth were taxed. It is in the last case that he pays the greatest price for his cloth; but all his additional payment is gained by the state. In the second case, he pays more than in the first, but all he pays in addition is not received by the state, it is an increased price caused by difficulty of production, which is incurred because the easiest means of production are taken away from us by being fettered with a tax.

CHAPTER XII

A LAND-TAX, levied in proportion to the rent of land, and vary-ing with every variation of rent, is in effect a tax on rent; and as such a tax will not apply to that land which yields no rent, nor to the produce of that capital which is employed on the land with a view to profit merely, and which never pays rent; it will not in any way affect the price of raw produce, but will fall wholly on the landlords. In no respect would such a tax differ from a tax on rent. But if a land-tax be imposed on all cultivated land, however moderate that tax may be, it will be a tax on produce, and will therefore raise the price of produce. If No. 3 be the land last cultivated, although it should pay no rent, it cannot, after the tax, be cultivated, and afford the general rate of profit, unless the price of produce rise to meet the tax. Either capital will be withheld from that employment until the price of corn shall have risen, in consequence of demand, sufficiently to afford the usual profit; or if already employed on such land, it will quit it, to seek a more advantageous em-ployment. The tax cannot be removed to the landlord, for by the supposition he receives no rent. Such a tax may be pro-portioned to the quality of the land and the abundance of its produce, and then it differs in no respect from tithes; or it may be a fixed tax per acre on all land cultivated, whatever its quality may be.

A land-tax of this latter description would be a very unequal tax, and would be contrary to one of the four maxims with regard to taxes in general, to which, according to Adam Smith, all taxes should conform. The four maxims are as follow:—

1. " The subjects of every state ought to contribute towards the support of the government, as nearly as possible in proportion to their respective abilities.
2. " The tax which each individual is bound to pay ought to be certain, and not arbitrary.
3. " Every tax ought to be levied at the time or in the manner in which it is most likely to be convenient for the contributor to pay it.

4. " Every tax ought to be so contrived as both to take out
    and to keep out of the pockets of the people as little
    as possible, over and above what it brings into the
    public treasury of the state."

An equal land-tax, imposed indiscriminately and without any
regard to the distinction of its quality, on all land cultivated,
will raise the price of corn in proportion to the tax paid by the
cultivator of the land of the worst quality. Lands of different
quality, with the employment of the same capital, will yield
very different quantities of raw produce. If on the land which
yields a thousand quarters of corn with a given capital a tax
of £100 be laid, corn will rise 2s. per quarter to compensate the
farmer for the tax. But with the same capital on land of a
better quality, 2000 quarters may be produced, which at 2s.
a quarter advance would give £200; the tax, however, bearing
equally on both lands, will be £100 on the better as well as on
the inferior, and consequently the consumer of corn will be
taxed, not only to pay the exigencies of the state, but also to
give to the cultivator of the better land £100 per annum during
the period of his lease, and afterwards to raise the rent of the
landlord to that amount. A tax of this description, then,
would be contrary to the fourth maxim of Adam Smith—it
would take out and keep out of the pockets of the people more
than what it brought into the treasury of the state. The taille
in France, before the Revolution, was a tax of this description;
those lands only were taxed which were held by an ignoble
tenure, the price of raw produce rose in proportion to the tax,
and therefore they whose lands were not taxed were benefited
by the increase of their rent. Taxes on raw produce, as well as
tithes, are free from this objection: they raise the price of raw
produce, but they take from each quality of land a contribution
in proportion to its actual produce, and not in proportion to
the produce of that which is the least productive.

From the peculiar view which Adam Smith took of rent, from
his not having observed that much capital is expended in every
country on the land for which no rent is paid, he concluded that
all taxes on the land, whether they were laid on the land itself
in the form of land-tax or tithes, or on the produce of the land,
or were taken from the profits of the farmer, were all invariably
paid by the landlord, and that he was in all cases the real contributor, although the tax was, in general, nominally advanced
by the tenant. " Taxes upon the produce of the land," he says,

" are in reality taxes upon the rent; and though they may be originally advanced by the farmer, are finally paid by the landlord. When a certain portion of the produce is to be paid away for a tax, the farmer computes as well as he can what the value of this portion is, one year with another, likely to amount to, and he makes a proportionable abatement in the rent which he agrees to pay to the landlord. There is no farmer who does not compute beforehand what the church-tithe, which is a land-tax of this kind, is, one year with another, likely to amount to." It is undoubtedly true that the farmer does calculate his probable outgoings of all descriptions when agreeing with his landlord for the rent of his farm; and if, for the tithe paid to the church, or for the tax on the produce of the land, he were not compensated by a rise in the relative value of the produce of his farm, he would naturally endeavour to deduct them from his rent. But this is precisely the question in dispute: whether he will eventually deduct them from his rent, or be compensated by a higher price of produce. For the reasons which have been already given, I cannot have the least doubt but that they would raise the price of produce, and consequently that Adam Smith has taken an incorrect view of this important question.

Dr. Smith's view of this subject is probably the reason why he has described " the tithe, and every other land-tax of this kind, under the appearance of perfect equality, as very unequal taxes; a certain portion of the produce being in different situations equivalent to a very different portion of the rent." I have endeavoured to show that such taxes do not fall with unequal weight on the different classes of farmers or landlords, as they are both compensated by the rise of raw produce, and only contribute to the tax in proportion as they are consumers of raw produce. Inasmuch indeed as wages, and through wages, the rate of profits are affected, landlords, instead of contributing their full share to such a tax, are the class peculiarly exempted. It is the profits of stock from which that portion of the tax is derived which falls on those labourers, who, from the insufficiency of their funds, are incapable of paying taxes; this portion is exclusively borne by all those whose income is derived from the employment of stock, and therefore it in no degree affects landlords.

It is not to be inferred from this view of tithes, and taxes on the land and its produce, that they do not discourage cultivation. Everything which raises the exchangeable value of commodities of any kind which are in very general demand tends to discourage both cultivation and production; but this is an evil inseparable from all taxation, and is not confined to the particular taxes of which we are now speaking.

This may be considered, indeed, as the unavoidable disadvantage attending all taxes received and expended by the state. Every new tax becomes a new charge on production, and raises natural price. A portion of the labour of the country which was before at the disposal of the contributor to the tax is placed at the disposal of the state, and cannot therefore be employed productively. This portion may become so large that sufficient surplus produce may not be left to stimulate the exertions of those who usually augment by their savings the capital of the state. Taxation has happily never yet in any free country been carried so far as constantly from year to year to diminish its capital. Such a state of taxation could not be long endured; or if endured, it would be constantly absorbing so much of the annual produce of the country as to occasion the most extensive scene of misery, famine, and depopulation.

" A land-tax," says Adam Smith, " which, like that of Great Britain, is assessed upon each district according to a certain invariable canon, though it should be equal at the time of its first establishment, necessarily becomes unequal in process of time, according to the unequal degrees of improvement or neglect in the cultivation of the different parts of the country. In England the valuation according to which the different counties and parishes were assessed to the land-tax by the 4th William and Mary was very unequal, even at its first establishment. This tax, therefore, so far offends against the first of the four maxims above mentioned. It is perfectly agreeable to the other three. It is perfectly certain. The time of payment for the tax being the same as that for the rent, is as convenient as it can be to the contributor. Though the landlord is in all cases the real contributor, the tax is commonly advanced by the tenant, to whom the landlord is obliged to allow it in the payment of the rent."

If the tax be shifted by the tenant not on the landlord but on the consumer, then if it be not unequal at first, it can never become so; for the price of produce has been at once raised in proportion to the tax, and will afterwards vary no more on that account. It may offend, if unequal, as I have attempted to show that it will, against the fourth maxim above mentioned, but it will not offend against the first. It may take more out of the pockets of the people than it brings into the public treasury of the state, but it will not fall unequally on any particular class of contributors. M. Say appears to me to have mistaken the nature and effects of the English land-tax, when he says, " Many persons attribute to this fixed valuation the great prosperity of English agriculture. That it has very much contributed to it there can be no doubt. But what should we say to a government which, addressing itself to a small trader, should hold this language: ' With a small capital you are carrying on a limited trade, and your direct contribution is in consequence very small. Borrow and accumulate capital; extend your trade, so that it may procure you immense profits; yet you shall never pay a greater contribution. Moreover, when your successors shall inherit your profits, and shall have further increased them, they shall not be valued higher to them than they are to you; and your successors shall not bear a greater portion of the public burdens.'

" Without doubt this would be a great encouragement given to manufacturers and trade; but would it be just? Could not their advancement be obtained at any other price? In England itself, has not manufacturing and commercial industry made even greater progress, since the same period, without being distinguished with so much partiality? A landlord by his assiduity, economy, and skill increases his annual revenue by 5000 francs. If the state claim of him the fifth part of his augmented income, will there not remain 4000 francs of increase to stimulate his further exertions? "

M. Say supposes, " A landlord by his assiduity, economy, and skill to increase his annual revenue by 5000 francs; " but a landlord has no means of employing his assiduity, economy, and skill on his land unless he farms it himself; and then it is in quality of capitalist and farmer that he makes the improvement, and not in quality of landlord. It is not conceivable that he could so augment the produce of his farm by any *peculiar* skill on his part, without first increasing the quantity of capital employed upon it. If he increased the capital, his larger revenue might bear the same proportion to his increased capital, as the revenue of all other farmers to their capitals.

If M. Say's suggestion were followed, and the state were to claim the fifth part of the augmented income of the farmer, it would be a partial tax on farmers, acting on their profits, and not affecting the profits of those in other employments. The tax would be paid by all lands, by those which yielded scantily as well as by those which yielded abundantly; and on some lands there could be no compensation for it by deduction from rent, for no rent is paid. A partial tax on profits never falls on the trade on which it is laid, for the trader will either quit his employment or remunerate himself for the tax. Now, those who pay no rent could be recompensed only by a rise in the price of produce, and thus would M. Say's proposed tax fall on the consumer, and not either on the landlord or farmer.

If the proposed tax were increased in proportion to the increased quantity or value of the gross produce obtained from the land, it would differ in nothing from tithes, and would equally be transferred to the consumer. Whether then it fell on the gross or on the net produce of land, it would be equally a tax on consumption, and would only affect the landlord and farmer in the same way as other taxes on raw produce.

If no tax whatever had been laid on the land, and the same sum had been raised by any other means, agriculture would have flourished at least as well as it has done; for it is impossible that any tax on land can be an *encouragement* to agriculture; a moderate tax may not, and probably does not, greatly prevent, but it cannot encourage production. The English government has held no such language as M. Say has supposed. It did not promise to exempt the agricultural class and their successors from all future taxation, and to raise the further supplies which the state might require from the other classes of society; it said only, " in this mode we will no further burthen the land; but we retain to ourselves the most perfect liberty of making you pay, under some other form, your full quota to the future exigencies of the state."

Speaking of taxes in kind, or a tax of a certain proportion of the produce, which is precisely the same as tithes, M. Say says, " This mode of taxation appears to be the most equitable; there is, however, none which is less so: it totally leaves out of consideration the advances made by the producer; it is proportioned by the gross, and not to the net revenue. Two agriculturists cultivate different kinds of raw produce: one cultivates corn on middling land, his expenses amounting annually on an average to 8000 francs; the raw produce from his lands sells for 12,000 francs; he has then a net revenue of 4000 francs.

" His neighbour has pasture or wood land, which brings in every year a like sum of 12,000 francs, but his expenses amount only to 2000 francs. He has therefore on an average a net revenue of 10,000 francs.

" A law ordains that a twelfth of the produce of all the fruits
of the earth be levied in kind, whatever they may be. From
the first is taken, in consequence of this law, corn of the value
of 1000 francs; and from the second, hay, cattle, or wood, of
the same value of 1000 francs. What has happened? From
the one, a quarter of his net income, 4000 francs, has been
taken; from the other, whose income was 10,000 francs, a tenth
only has been taken. Income is the net profit which remains
after replacing the capital exactly in its former state. Has a
merchant an income equal to all the sales which he makes in
the course of a year; certainly not; his income only amounts
to the excess of his sales above his advances, and it is on this
excess only that taxes on income should fall."

M. Say's error in the above passage lies in supposing that
because the value of the produce of one of these two farms, after
reinstating the capital, is greater than the value of the produce
of the other, on that account the net income of the cultivators
will differ by the same amount. The net income of the landlords and tenants together of the wood land may be much
greater than the net income of the landlords and tenants of the
corn land; but it is on account of the difference of rent, and not
on account of the difference in the rate of profit. M. Say has
wholly omitted the consideration of the different amount of
rent which these cultivators would have to pay. There cannot
be two rates of profit in the same employment, and therefore
when the value of produce is in different proportions to capital,
it is the rent which will differ, and not the profit. Upon what
pretence would one man, with a capital of 2000 francs, be
allowed to obtain a net profit of 10,000 francs from its employment, whilst another, with a capital of 8000 francs, would only
obtain 4000 francs? Let M. Say make a due allowance for rent;
let him further allow for the effect which such a tax would have
on the prices of these different kinds of raw produce, and he will
then perceive that it is not an unequal tax, and, further, that
the producers themselves will no otherwise contribute to it
than any other class of consumers.

CHAPTER XIII

TAXES ON GOLD

THE rise in the price of commodities, in consequence of taxation or of difficulty of production, will in all cases ultimately ensue; but the duration of the interval before the market price will conform to the natural price must depend on the nature of the commodity, and on the facility with which it can be reduced in quantity. If the quantity of the commodity taxed could not be diminished, if the capital of the farmer or of the hatter, for instance, could not be withdrawn to other employments, it would be of no consequence that their profits were reduced below the general level by means of a tax; unless the demand for their commodities should increase, they would never be able to elevate the market price of corn and of hats up to their increased natural price. Their threats to leave their employments, and remove their capitals to more favoured trades, would be treated as an idle menace which could not be carried into effect; and consequently the price would not be raised by diminished production. Commodities, however, of all descriptions, can be reduced in quantity, and capital can be removed from trades which are less profitable to those which are more so, but with different degrees of rapidity. In proportion as the supply of a particular commodity can be more easily reduced, without inconvenience to the producer, the price of it will more quickly rise after the difficulty of its production has been increased by taxation, or by any other means. Corn being a commodity indispensably necessary to every one, little effect will be produced on the demand for it in consequence of a tax, and therefore the supply would not probably be long excessive, even if the producers had great difficulty in removing their capitals from the land. For this reason, the price of corn will speedily be raised by taxation, and the farmer will be enabled to transfer the tax from himself to the consumer.

If the mines which supply us with gold were in this country, and if gold were taxed, it could not rise in relative value to other things till its quantity were reduced. This would be more particularly the case if gold were used exclusively for money.

It is true that the least productive mines, those which paid no rent, could no longer be worked, as they could not afford the general rate of profits till the relative value of gold rose by a sum equal to the tax. The quantity of gold, and, therefore, the quantity of money, would be slowly reduced: it would be a little diminished in one year, a little more in another, and finally its value would be raised in proportion to the tax; but, in the interval, the proprietors or holders, as they would pay the tax, would be the sufferers, and not those who used money. If out of every 1000 quarters of wheat in the country, and every 1000 produced in future, government should exact 100 quarters as a tax, the remaining 900 quarters would exchange for the same quantity of other commodities that 1000 did before; but if the same thing took place with respect to gold, if of every £1000 money now in the country, or in future to be brought into it, government could exact £100 as a tax, the remaining £900 would purchase very little more than £900 purchased before. The tax would fall upon him whose property consisted of money, and would continue to do so till its quantity were reduced in proportion to the increased cost of its production caused by the tax.

This, perhaps, would be more particularly the case with respect to a metal used for money than any other commodity; because the demand for money is not for a definite quantity, as is the demand for clothes, or for food. The demand for money is regulated entirely by its value, and its value by its quantity. If gold were of double the value, half the quantity would perform the same functions in circulation, and if it were of half the value, double the quantity would be required. If the market value of corn be increased one-tenth by taxation, or by difficulty of production, it is doubtful whether any effect whatever would be produced on the quantity consumed, because every man's want is for a definite quantity, and, therefore, if he has the means of purchasing, he will continue to consume as before: but for money, the demand is exactly proportioned to its value. No man could consume twice the quantity of corn which is usually necessary for his support, but every man purchasing and selling only the same quantity of goods may be obliged to employ twice, thrice, or any number of times the same quantity of money.

The argument which I have just been using applies only to those states of society in which the precious metals are used for money, and where paper credit is not established. The metal gold, like all other commodities, has its value in the market ultimately regulated by the comparative facility or difficulty of producing it; and although, from its durable nature, and from the difficulty of reducing its quantity, it does not readily bend to variations in its market value, yet that difficulty is much increased from the circumstance of its being used as money. If the quantity of gold in the market for the purpose of commerce only were 10,000 ounces, and the consumption in our manufactures were 2000 ounces annually, it might be raised one-fourth or 25 per cent. in its value in one year by withholding the annual supply; but if, in consequence of its being used as money, the quantity employed were 100,000 ounces, it would not be raised one-fourth in value in less than ten years. As money made of paper may be readily reduced in quantity, its value, though its standard were gold, would be increased as rapidly as that of the metal itself would be increased, if the metal, by forming a very small part of the circulation, had a very slight connection with money.

If gold were the produce of one country only, and it were used universally for money, a very considerable tax might be imposed on it, which would not fall on any country, except in proportion as they used it in manufactures and for utensils; upon that portion which was used for money, though a large tax might be received, nobody would pay it. This is a quality peculiar to money. All other commodities of which there exists a limited quantity, and which cannot be increased by competition, are dependent for their value on the tastes, the caprice, and the power of purchasers; but money is a commodity which no country has any wish or necessity to increase: no more advantage results from using twenty millions than from using ten millions of currency. A country might have a monopoly of silk, or of wine, and yet the prices of silks and wine might fall, because from caprice, or fashion, or taste, cloth and brandy might be preferred and substituted; the same effect might in a degree take place with gold, as far as its use is confined to manufactures: but while money is the general medium of exchange, the demand for it is never a matter of choice, but always of necessity: you must take it in exchange for your goods, and, therefore, there are no limits to the quantity which may be forced on you by foreign trade if it fall in value; and no reduction to which you must not submit if it rise. You may, indeed, substitute paper money, but by this you do not and cannot lessen the quantity of money, for that is regulated by the value of the standard for which it is exchangeable; it is only by the rise of the price of commodities that you can prevent them from being exported from a country where they are purchased with little money, to a country where they can be sold for more, and this rise can only be effected by an importation of metallic money from abroad, or by the creation or addition of paper money at home. If, then, the King of Spain, supposing him to be in exclusive possession of the mines, and gold alone to be used for money, were to lay a considerable tax on gold, he would very much raise its natural value; and as its market value in Europe is ultimately regulated by its natural value in Spanish America, more commodities would be given by Europe for a given quantity of gold. But the same quantity of gold would not be produced in America, as its value would only be increased in proportion to the diminution of quantity consequent on its increased cost of production. No more goods, then, would be obtained in America in exchange for all their gold exported than before; and it may be asked where then would be the benefit to Spain and her colonies? The benefit would be this, that if less gold were produced, less capital would be employed in producing it; the same value of goods from Europe would be imported by the employment of the smaller capital that was before obtained by the employment of the larger; and, therefore, all the productions obtained by the employment of the capital withdrawn from the mines would be a benefit which Spain would derive from the imposition of the tax, and which she could not obtain in such abundance, or with such certainty, by possessing the monopoly of any other commodity whatever. From such a tax, as far as money was concerned, the nations of Europe would suffer no injury whatever; they would have the same quantity of goods, and consequently the same means of enjoyment as before, but these goods would be circulated with a less quantity, because a more valuable money.

If in consequence of the tax only one-tenth of the present quantity of gold were obtained from the mines, that tenth would be of equal value with the ten tenths now produced. But the King of Spain is not exclusively in possession of the mines of the precious metals; and if he were, his advantage from their possession, and the power of taxation, would be very much reduced by the limitation of demand and consumption in Europe, in consequence of the universal substitution, in a greater or less degree, of paper money. The agreement of the market and natural prices of all commodities depends at all times on the facility with which the supply can be increased or diminished. In the case of gold, houses, and labour, as well as many other things, this effect cannot, under some circumstances, be speedily produced. But it is different with those commodities which are consumed and reproduced from year to year, such as hats, shoes, corn, and cloth; they may be reduced, if necessary, and the interval cannot be long before the supply is contracted in proportion to the increased charge of producing them.

A tax on raw produce from the surface of the earth will, as we have seen, fall on the consumer, and will in no way affect rent; unless by diminishing the funds for the maintenance of labour it lowers wages, reduces the population, and diminishes the demand for corn. But a tax on the produce of gold mines must, by enhancing the value of that metal, necessarily reduce the demand for it, and must therefore necessarily displace capital from the employment to which it was applied. Notwithstanding, then, that Spain would derive all the benefits which I have stated from a tax on gold, the proprietors of those mines from which capital was withdrawn would lose all their rent. This would be a loss to individuals, but not a national loss; rent being not a creation, but merely a transfer of wealth: the King of Spain, and the proprietors of the mines which continued to be worked, would together receive, not only all that the liberated capital produced, but all that the other proprietors lost.

Suppose the mines of the 1st, 2nd, and 3rd quality to be worked, and to produce respectively 100, 80, and 70 pounds' weight of gold, and therefore the rent of No. 1 to be thirty pounds, and that of No. 2 ten pounds. Suppose, now, the tax to be seventy pounds of gold per annum on each mine worked; and consequently that No. 1 alone could be profitably worked, it is evident that all rent would immediately disappear. Before the imposition of the tax, out of the 100 pounds produced on No. 1, a rent was paid of thirty pounds, and the worker of the mine retained seventy, a sum equal to the produce of the least productive mine. The value, then, of what remains to the capitalist of the mine No. 1 must be the same as before, or he would not obtain the common profits of stock; and, consequently, after paying seventy out of his 100 pounds for tax, the value of the remaining thirty must be as great as the value of seventy was before, and therefore the value of the whole hundred as great as 233 pounds before. Its value might be higher, but it could not be lower, or even this mine would cease to be worked. Being a monopolised commodity, it could exceed its natural value, and then it would pay a rent equal to that excess; but no funds would be employed in the mine if it were below this value. In return for one-third of the labour and capital employed in the mines, Spain would obtain as much gold as would exchange for the same, or very nearly the same, quantity of commodities as before. She would be richer by the produce of the two-thirds liberated from the mines. If the value of the 100 pounds of gold should be equal to that of the 250 pounds extracted before, the King of Spain's portion, his seventy pounds would be equal to 175 at the former value: a small part of the king's tax only would fall on his own subjects, the greater part being obtained by the better distribution of capital.

The account of Spain would stand thus:—

FORMERLY PRODUCED

Gold, 250 pounds, of the value of (suppose) . .	10,000 yards of cloth.

NOW PRODUCED

By the two capitalists who quitted the mines, the same value as 140 pounds of gold formerly exchanged for; equal to	5600 yards of cloth.
By the capitalist who works the mine, No. 1, thirty pounds of gold, increased in value, as 1 to 2½, and therefore now of the value of	3000 yards of cloth.
Tax to the king, seventy pounds, increased also in value as 1 to 2½, and therefore now of the value of	7000 yards of cloth.
	15,600

Of the 7000 received by the king, the people of Spain would contribute only 1400, and 5600 would be pure gain, effected by the liberated capital.

If the tax, instead of being a fixed sum per mine worked, were a certain portion of its produce, the quantity would not be immediately reduced in consequence. If a half, a fourth, or a third of each mine were taken for the tax, it would nevertheless be the interest of the proprietors to make their mines yield as abundantly as before; but if the quantity were not reduced, but only a part of it transferred from the proprietor to the king, its value would not rise; the tax would fall on the people of the colonies, and no advantage would be gained. A tax of this kind would have the effect that Adam Smith supposes taxes on raw produce would have on the rent of land—it would fall entirely on the rent of the mine. If pushed a little further, indeed, the tax would not only absorb the whole rent, but would deprive the worker of the mine of the common profits of stock, and he would consequently withdraw his capital from the production of gold. If still further extended, the rent of still better mines would be absorbed, and capital would be further withdrawn; and thus the quantity would be continually reduced, and its value raised, and the same effects would take place as we have already pointed out; a part of the tax would be paid by the people of the Spanish colonies, and the other part would be a new creation of produce, by increasing the power of the instrument used as a medium of exchange.

Taxes on gold are of two kinds, one on the actual quantity of gold in circulation, the other on the quantity that is annually produced from the mines. Both have a tendency to reduce the quantity and to raise the value of gold; but by neither will its value be raised till the quantity is reduced, and therefore such taxes will fall for a time, until the supply is diminished, on the proprietors of money, but ultimately that part which will permanently fall on the community will be paid by the owner of the mine in the reduction of rent, and by the purchasers of that portion of gold which is used as a commodity contributing to the enjoyments of mankind, and not set apart exclusively for a circulating medium.

CHAPTER XIV

THERE are also other commodities besides gold which cannot be speedily reduced in quantity; any tax on which will therefore fall on the proprietor if the increase of price should lessen the demand.

Taxes on houses are of this description; though laid on the occupier, they will frequently fall by a diminution of rent on the landlord. The produce of the land is consumed and reproduced from year to year, and so are many other commodities; as they may therefore be speedily brought to a level with the demand, they cannot long exceed their natural price. But as a tax on houses may be considered in the light of an additional rent paid by the tenant, its tendency will be to diminish the demand for houses of the same annual rent without diminishing their supply. Rent will therefore fall, and a part of the tax that will be paid indirectly by the landlord.

"The rent of a house," says Adam Smith, "may be distinguished into two parts, of which the one may very properly be called the building rent, the other is commonly called the ground rent. The building rent is the interest or profit of the capital expended in building the house. In order to put the trade of a builder upon a level with other trades, it is necessary that this rent should be sufficient first to pay the same interest which he would have got for his capital if he had lent it upon good security; and, secondly, to keep the house in constant repair, or, what comes to the same thing, to replace within a certain term of years the capital which had been employed in building it." "If, in proportion to the interest of money, the trade of the builder affords at any time a much greater profit than this, it will soon draw so much capital from other trades as will reduce the profit to its proper level. If it affords at any time much less than this, other trades will soon draw so much capital from it as will again raise that profit. Whatever part of the whole rent of a house is over and above what is sufficient for affording this reasonable profit, naturally goes to the ground rent; and where the owner of the ground, and the owner of the building, are two different persons, it is in most cases completely paid to the former. In country houses, at a distance from any great town, where there is a plentiful choice of ground, the ground rent is scarcely anything, or no more than what the space upon which the house stands would pay employed in agriculture. In country villas, in the neighbourhood of some great town, it is sometimes a good deal higher, and the peculiar conveniency, or beauty of situation, is there frequently very highly paid for. Ground rents are generally highest in the capital, and in those particular parts of it where there happens to be the greatest demand for houses, whatever be the reason for that demand, whether for trade and business, for pleasure and society, or for mere vanity and fashion." A tax on the rent of houses may either fall on the occupier, on the ground landlord, or on the building landlord. In ordinary cases it may be presumed that the whole tax would be paid, both immediately and finally, by the occupier.

If the tax be moderate, and the circumstances of the country such that it is either stationary or advancing, there would be little motive for the occupier of a house to content himself with one of a worse description. But if the tax be high, or any other circumstances should diminish the demand for houses, the landlord's income would fall, for the occupier would be partly compensated for the tax by a diminution of rent. It is, however, difficult to say in what proportions that part of the tax, which was saved by the occupier by a fall of rent, would fall on the building rent and the ground rent. It is probable that, in the first instance, both would be affected; but as houses are, though slowly, yet certainly perishable, and as no more would be built till the profits of the builder were restored to the general level, building rent would, after an interval, be restored to its natural price. As the builder receives rent only whilst the building endures, he could pay no part of the tax, under the most disastrous circumstances, for any longer period.

The payment of this tax, then, would ultimately fall on the occupier and ground landlord, but, " in what proportion this final payment would be divided between them," says Adam Smith, " it is not perhaps very easy to ascertain. The division would probably be very different in different circumstances, and a tax of this kind might, according to those different circumstances, affect very unequally both the inhabitant of the house and the owner of the ground." [1]

---

[1] Book v. chap. ii.

Adam Smith considers ground rents as peculiarly fit subjects for taxation. "Both ground rents and the ordinary rent of land," he says, "are a species of revenue, which the owner in many cases enjoys without any care or attention of his own. Though a part of this revenue should be taken from him, in order to defray the expenses of the state, no discouragement will thereby be given to any sort of industry. The annual produce of the land and labour of the society, the real wealth and revenue of the great body of the people, might be the same after such a tax as before. Ground rents and the ordinary rent of land are, therefore, perhaps, the species of revenue which can best bear to have a peculiar tax imposed upon them." It must be admitted that the effects of these taxes would be such as Adam Smith has described; but it would surely be very unjust to tax exclusively the revenue of any particular class of a community. The burdens of the state should be borne by all in proportion to their means: this is one of the four maxims mentioned by Adam Smith which should govern all taxation. Rent often belongs to those who, after many years of toil, have realised their gains and expended their fortunes in the purchase of land or houses; and it certainly would be an infringement of that principle which should ever be held sacred, the security of property, to subject it to unequal taxation. It is to be lamented that the duty by stamps, with which the transfer of landed property is loaded, materially impedes the conveyance of it into those hands where it would probably be made most productive. And if it be considered that land, regarded as a fit subject for exclusive taxation, would not only be reduced in price, to compensate for the risk of that taxation, but in proportion to the indefinite nature and uncertain value of the risk would become a fit subject for speculations, partaking more of the nature of gambling than of sober trade, it will appear probable that the hands into which land would in that case be most apt to fall would be the hands of those who possess more of the qualities of the gambler than of the qualities of the sober-minded proprietor, who is likely to employ his land to the greatest advantage.

CHAPTER XV

TAXES ON PROFITS

TAXES on those commodities which are generally denominated luxuries fall on those only who make use of them. A tax on wine is paid by the consumer of wine. A tax on pleasure horses, or on coaches, is paid by those who provide for themselves such enjoyments, and in exact proportion as they provide them. But taxes on necessaries do not affect the consumers of necessaries in proportion to the quantity that may be consumed by them, but often in a much higher proportion. A tax on corn, we have observed, not only affects a manufacturer in the proportion that he and his family may consume corn, but it alters the rate of profits of stock, and therefore also affects his income. Whatever raises the wages of labour, lowers the profits of stock; therefore every tax on any commodity consumed by the labourer has a tendency to lower the rate of profits.

A tax on hats will raise the price of hats; a tax on shoes, the price of shoes; if this were not the case, the tax would be finally paid by the manufacturer; his profits would be reduced below the general level, and he would quit his trade. A partial tax on profits will raise the price of the commodity on which it falls: a tax, for example, on the profits of the hatter would raise the price of hats; for if his profits were taxed, and not those of any other trade, his profits, unless he raised the price of his hats, would be below the general rate of profits, and he would quit his employment for another.

In the same manner, a tax on the profits of the farmer would raise the price of corn; a tax on the profits of the clothier, the price of cloth; and if a tax in proportion to profits were laid on all trades, every commodity would be raised in price. But if the mine which supplied us with the standard of our money were in this country, and the profits of the miner were also taxed, the price of no commodity would rise, each man would give an equal proportion of his income, and everything would be as before.

If money be not taxed, and therefore be permitted to preserve its value, whilst everything else is taxed and is raised in value, the hatter, the farmer, and clothier, each employing the same capitals, and obtaining the same profits, will pay the same amount of tax. If the tax be £100, the hats, the cloth, and the corn will each be increased in value £100. If the hatter gains by his hats £1100, instead of £1000, he will pay £100 to government for the tax; and therefore will still have £1000 to lay out on goods for his own consumption. But as the cloth, corn, and all other commodities will be raised in price from the same cause, he will not obtain more for his £1000 than he before obtained for £910, and thus will he contribute by his diminished expenditure to the exigencies of the state; he will, by the payment of the tax, have placed a portion of the produce of the land and labour of the country at the disposal of government, instead of using that portion himself. If, instead of expending his £1000, he adds it to his capital, he will find in the rise of wages, and in the increased cost of the raw material and machinery, that his saving of £1000 does not amount to more than a saving of £910 amounted to before.

If money be taxed, or if by any other cause its value be altered, and all commodities remain precisely at the same price as before, the profits of the manufacturer and farmer will also be the same as before, they will continue to be £1000; and as they will each have to pay £100 to government, they will retain only £900, which will give them a less command over the produce of the land and labour of the country, whether they expend it in productive or unproductive labour. Precisely what they lose, government will gain. In the first case, the contributor to the tax would, for £1000, have as great a quantity of goods as he before had for £910; in the second, he would have only as much as he before had for £900, for the price of goods would remain unaltered, and he would have only £900 to expend. This proceeds from the difference in the amount of the tax; in the first case, it is only an eleventh of his income; in the second, it is a tenth; money in the two cases being of a different value.

But although, if money be not taxed, and do not alter in value, all commodities will rise in price, they will not rise in the same proportion; they will not after the tax bear the same relative value to each other which they did before the tax. In a former part of this work we discussed the effects of the division of capital into fixed and circulating, or rather into durable and perishable capital, on the prices of commodities. We showed that two manufacturers might employ precisely the same amount of capital, and might derive from it precisely the same amount of profits, but that they would sell their commodities for very different sums of money, according as the capitals they employed were rapidly, or slowly, consumed and reproduced. The one might sell his goods for £4000, the other for £10,000, and they might both employ £10,000 of capital, and obtain 20 per cent. profit, or £2000. The capital of one might consist, for example, of £2000 circulating capital, to be reproduced, and £8000 fixed, in buildings and machinery; the capital of the other, on the contrary, might consist of £8000 of circulating, and of only £2000 fixed capital in machinery and buildings. Now, if each of these persons were to be taxed 10 per cent. on his income, or £200, the one to make his business yield him the general rate of profit must raise his goods from £10,000 to £10,200; the other would also be obliged to raise the price of his goods from £4000 to £4200. Before the tax, the goods sold by one of these manufacturers were $2\frac{1}{2}$ times more valuable than the goods of the other; after the tax they will be 2.42 times more valuable: the one kind will have risen two per cent.: the other five per cent.: consequently a tax upon income, whilst money continued unaltered in value, would alter the relative prices and value of commodities. This would be true also if the tax, instead of being laid on the profits, were laid on the commodities themselves: provided they were taxed in proportion to the value of the capital employed on their production, they would rise equally, whatever might be their value, and therefore they would not preserve the same proportion as before. A commodity which rose from ten to eleven thousand pounds would not bear the same relation as before to another which rose from £2000 to £3000. If, under these circumstances, money rose in value, from whatever cause it might proceed, it would not affect the prices of commodities in the same proportion. The same cause which would lower the price of one from £10,200 to £10,000 or less than two per cent., would lower the price of the other from £4200 to £4000 or $4\frac{3}{4}$ per cent. If they fell in any different proportion, profits would not be equal; for to make them equal, when the price of the first commodity was £10,000, the price of the second should be £4000; and when the price of the first was £10,200, the price of the other should be £4200.

The consideration of this fact will lead to the understanding of a very important principle, which, I believe, has never been adverted to. It is this : that in a country where no taxation subsists, the alteration in the value of money arising from scarcity or abundance will operate in an equal proportion on the prices of all commodities; that if a commodity of £1000 value rise to £1200, or fall to £800, a commodity of £10,000 value will rise to £12,000 or fall to £8000; but in a country where prices are artificially raised by taxation, the abundance of money from an influx, or the exportation and consequent scarcity of it from foreign demand, will not operate in the same proportion on the prices of all commodities; some it will raise or lower 5, 6, or 12 per cent., others 3, 4, or 7 per cent. If a country were not taxed, and money should fall in value, its abundance in every market would produce similar effects in each. If meat rose 20 per cent., bread, beer, shoes, labour, and every commodity would also rise 20 per cent.; it is necessary they should do so, to secure to each trade the same rate of profits. But this is no longer true when any of these commodities is taxed; if, in that case, they should all rise in proportion to the fall in the value of money, profits would be rendered unequal; in the case of the commodities taxed, profits would be raised above the general level, and capital would be removed from one employment to another, till an equilibrium of profits was restored, which could only be after the relative prices were altered.

Will not this principle account for the different effects, which it was remarked were produced on the prices of commodities from the altered value of money during the bank-restriction? It was objected to those who contended that the currency was at that period depreciated, from the too great abundance of the paper circulation, that, if that were the fact, all commodities ought to have risen in the same proportion; but it was found that many had varied considerably more than others, and thence it was inferred that the rise of prices was owing to something affecting the value of commodities, and not to any alteration in the value of the currency. It appears, however, as we have just seen, that in a country where commodities are taxed, they will not all vary in price in the same proportion, either in consequence of a rise or of a fall in the value of currency.

If the profits of all trades were taxed, excepting the profits of the farmer, all goods would rise in money value, excepting raw produce. The farmer would have the same corn income as before, and would sell his corn also for the same money price; but as he would be obliged to pay an additional price for all the commodities, except corn, which he consumed, it would be to him a tax on expenditure. Nor would he be relieved from this tax by an alteration in the value of money, for an alteration in the value of money might sink all the taxed commodities to their former price, but the untaxed one would sink below its former level; and, therefore, though the farmer would purchase his commodities at the same price as before, he would have less money with which to purchase them.

The landlord, too, would be precisely in the same situation; he would have the same corn, and the same money-rent as before, if all commodities rose in price and money remained at the same value; and he would have the same corn, but a less money-rent, if all commodities remained at the same price: so that in either case, though his income were not directly taxed, he would indirectly contribute towards the money raised.

But suppose the profits of the farmer to be also taxed, he then would be in the same situation as other traders: his raw produce would rise, so that he would have the same money revenue, after paying the tax, but he would pay an additional price for all the commodities he consumed, raw produce included.

His landlord, however, would be differently situated; he would be benefited by the tax on his tenant's profits, as he would be compensated for the additional price at which he would purchase his manufactured commodities, if they rose in price; and he would have the same money revenue, if, in consequence of a rise in the value of money, commodities sold at their former price. A tax on the profits of the farmer is not a tax proportioned to the gross produce of the land, but to its net produce, after the payment of rent, wages, and all other charges. As the cultivators of the different kinds of land, No. 1, 2, and 3, employ precisely the same capitals, they will get precisely the same profits, whatever may be the quantity of gross produce which one may obtain more than the other; and consequently they will be all taxed alike. Suppose the gross produce of the land of the quality of No. 1 to be 180 qrs., that of No. 2, 170 qrs., and of No. 3, 160, and each to be taxed 10 quarters, the difference between the produce of No. 1, No. 2, and No. 3, after paying the tax, will be the same as before; for if No. 1 be reduced to 170, No. 2 to 160, and No. 3 to 150 qrs., the difference between 3 and 1 will be as before, 20 qrs.; and of No. 3 and No. 2, 10 qrs. If, after the tax, the prices of corn and of every other commodity should remain the same as before, money rent, as well as corn rent, would continue unaltered; but if the price of corn and every other commodity should rise in consequence of the tax, money rent will also rise in the same proportion. If the price of corn were £4 per quarter, the rent of No. 1 would be £80, and that of No. 2, £40; but if corn rose five per cent., or to £4 4s., rent would also rise five per cent., for twenty quarters of corn would then be worth £84, and ten quarters £42; so that in every case the landlord will be unaffected by such a tax. A tax on the profits of stock always leaves corn rent unaltered, and therefore money rent varies with the price of corn; but a tax on raw produce, or tithes, never leaves corn rent unaltered, but generally leaves money rent the same as before. In another part of this work I have observed that if a land-tax of the same money amount were laid on every kind of land in cultivation, without any allowance for difference of fertility, it would be very unequal in its operation, as it would be a profit to the landlord of the more fertile lands. It would raise the price of corn in proportion to the burden borne by the farmer of the worst land; but this additional price being obtained for the greater quantity of produce yielded by the better land, farmers of such land would be benefited during their leases, and afterwards the advantage would go to the landlord in the form of an increase of rent. The effect of an equal tax on the *profits* of the farmer is precisely the same; it raises the money rent of the landlords if money retains the same value; but as the profits of all other trades are taxed as well as those of the farmer, and consequently the prices of all goods, as well as corn, are raised, the landlord loses as much by the increased money price of the goods and corn on which his rent is expended, as he gains by the rise of his rent. If money should rise in value, and all things should, after a tax on the profits of stock, fall to their former prices, rent also would be the same as before. The landlord would receive the same money rent, and would obtain all the commodities on which it was expended at their former price; so that under all circumstances he would continue untaxed.[1]

This circumstance is curious. By taxing the profits of the farmer you do not burthen him more than if you exempted his profits from the tax, and the landlord has a decided interest that his tenants' profits should be taxed, as it is only on that condition that he himself continues really untaxed.

A tax on the profits of capital would also affect the stock-

[1] That the profits of the farmer only should be taxed, and not the profits of any other capitalist, would be highly beneficial to landlords. It would, in fact, be a tax on the consumers of raw produce, partly for the benefit of the state, and partly for the benefit of landlords.

holder, if all commodities were to rise in proportion to the tax although his dividends continued untaxed; but if, from the alteration in the value of money, all commodities were to sink to their former price, the stock-holder would pay nothing towards the tax; he would purchase all his commodities at the same price, but would still receive the same money dividend.

If it be agreed that by taxing the profits of one manufacturer only, the price of his goods would rise, to put him on an equality with all other manufacturers; and that by taxing the profits of two manufacturers the prices of two descriptions of goods must rise, I do not see how it can be disputed that by taxing the profits of all manufacturers the prices of all goods would rise, provided the mine which supplied us with money were in this country and continued untaxed. But as money, or the standard of money, is a commodity imported from abroad, the prices of all goods could not rise; for such an effect could not take place without an additional quantity of money,[1] which could not be obtained in exchange for dear goods, as was shown in page 60. If, however, such a rise could take place, it could not be permanent, for it would have a powerful influence on foreign trade. In return for commodities imported, those dear goods could not be exported, and therefore we should for a time continue to buy, although we ceased to sell; and should export money, or bullion, till the relative prices of commodities were nearly the same as before. It appears to me absolutely certain that a well regulated tax on profits would ultimately restore commodities, both of home and foreign manufacture, to the same money price which they bore before the tax was imposed.

As taxes on raw produce, tithes, taxes on wages, and on

[1] On further consideration, I doubt whether any more money would be required to circulate the same quantity of commodities if their prices be raised by taxation and not by difficulty of production. Suppose 100,000 quarters of corn to be sold in a certain district, and in a certain time, at £4 per quarter, and that in consequence of a direct tax of 8s. per quarter, corn rises to £4 8s., the same quantity of money, I think, and no more, would be required to circulate this corn at the increased price. If I before purchased 11 quarters at £4, and, in consequence of the tax, am obliged to reduce my consumption to 10 quarters, I shall not require more money, for in all cases I shall pay £44 for my corn. The public would, in fact, consume one-eleventh less, and this quantity would be consumed by government. The money necessary to purchase it would be derived from the 8s. per quarter, to be received from the farmers in the shape of a tax, but the amount levied would at the same time be paid to them for their corn; therefore the tax is in fact a tax in kind, and does not make it necessary that any more money should be used, or, if any, so little that the quantity may be safely neglected.

the necessaries of the labourer will, by raising wages, lower profits, they will all, though not in an equal degree, be attended with the same effects.

The discovery of machinery, which materially improves home manufactures, always tends to raise the relative value of money, and therefore to encourage its importation. All taxation, all increased impediments, either to the manufacturer or the grower of commodities, tend, on the contrary, to lower the relative value of money, and therefore to encourage its exportation.

CHAPTER XVI

TAXES ON WAGES

TAXES on wages will raise wages, and therefore will diminish the rate of the profits of stock. We have already seen that a tax on necessaries will raise their prices, and will be followed by a rise of wages. The only difference between a tax on necessaries and a tax on wages is, that the former will necessarily be accompanied by a rise in the price of necessaries, but the latter will not; towards a tax on wages, consequently, neither the stockholder, the landlord, nor any other class but the employers of labour will contribute. A tax on wages is wholly a tax on profits; a tax on necessaries is partly a tax on profits and partly a tax on rich consumers. The ultimate effects which will result from such taxes, then, are precisely the same as those which result from a direct tax on profits.

"The wages of the inferior classes of workmen," says Adam Smith, "I have endeavoured to show in the first book, are everywhere necessarily regulated by two different circumstances—the demand for labour and the ordinary or average price of provisions. The demand for labour, according as it happens to be either increasing, stationary, or declining, or to require an increasing, stationary, or declining population, regulates the subsistence of the labourer, and determines in what degree it shall be either liberal, moderate, or scanty. The *ordinary or average* price of provisions determines the quantity of money which must be paid to the workmen, in order to enable him, one year with another, to purchase this liberal, moderate, or scanty subsistence. While the demand for labour and the price of provisions, therefore, remain the same, a direct tax upon the wages of labour can have no other effect than to raise them somewhat higher than the tax."

To the proposition, as it is here advanced by Dr. Smith, Mr. Buchanan offers two objections. First, he denies that the money wages of labour are regulated by the price of provisions; and secondly, he denies that a tax on the wages of labour would raise the price of labour. On the first point Mr. Buchanan's argument is as follows, page 59: "The wages of labour, it has already been remarked, consist not in money, but in what money purchases, namely, provisions and other necessaries; and the allowance of the labourer out of the common stock will always be in proportion to the supply. Where provisions are *cheap and abundant*, his share will be the larger; and where they are *scarce and dear*, it will be the less. His wages will always give him his just share, and they cannot give him more. It is an opinion, indeed, adopted by Dr. Smith and most other writers, that the money price of labour is regulated by the money price of provisions, and that, when provisions rise in price, wages rise in proportion. But it is clear that the price of labour has no necessary connection with the price of food, since it depends entirely on the supply of labourers compared with the demand. Besides, it is to be observed that the high price of provisions is a certain indication of a deficient supply, and arises in the natural course of things for the purpose of retarding the consumption. A smaller supply of food, shared among the same number of consumers, will evidently leave a smaller portion to each, and the labourer must bear his share of the common want. To distribute this burden equally, and to prevent the labourer from consuming subsistence so freely as before, the price rises. But wages, it seems, must rise along with it, that he may still use the same quantity of a scarcer commodity; and thus nature is represented as counteracting her own purposes;—first, raising the price of food to diminish the consumption, and afterwards raising wages to give the labourer the same supply as before."

In this argument of Mr. Buchanan, there appears to me to be a great mixture of truth and error. Because a high price of provisions is sometimes occasioned by a deficient supply, Mr. Buchanan assumes it as a certain indication of deficient supply. He attributes to one cause exclusively that which may arise from many. It is undoubtedly true that, in the case of a deficient supply, a smaller quantity will be shared among the same number of consumers, and a smaller portion will fall to each. To distribute this privation equally, and to prevent the labourer from consuming subsistence so freely as before, the price rises. It must, therefore, be conceded to Mr. Buchanan that any rise in the price of provisions occasioned by a deficient supply will not necessarily raise the money wages of labour as the consumption must be retarded, which can only be effected by diminishing the power of the consumers to purchase. But, because the price of provisions is raised by a deficient supply, we are by no means warranted in concluding, as Mr. Buchanan appears to do, that there may not be an abundant supply with a high price; not a high price with regard to money only, but with regard to all other things.

The natural price of commodities, which always ultimately governs their market price, depends on the facility of production; but the quantity produced is not in proportion to that facility. Although the lands which are now taken into cultivation are much inferior to the lands in cultivation three centuries ago, and therefore the difficulty of production is increased, who can entertain any doubt but that the quantity produced now very far exceeds the quantity then produced? Not only is a high price compatible with an increased supply, but it rarely fails to accompany it. If, then, in consequence of taxation, or of difficulty of production, the price of provisions be raised and the quantity be not diminished, the money wages of labour will rise; for, as Mr. Buchanan has justly observed, "The wages of labour consist not in money, but in what money purchases, namely, provisions and other necessaries; and the allowance of the labourer out of the common stock will always be in proportion to the supply."

With respect to the second point, whether a tax on the wages of labour would raise the price of labour, Mr. Buchanan says, "After the labourer has received the fair recompense of his labour, how can he have recourse on his employer for what he is afterwards compelled to pay away in taxes? There is no law or principle in human affairs to warrant such a conclusion. After the labourer has received his wages, they are in his own keeping, and he must, as far as he is able, bear the burden of whatever exactions he may ever afterwards be exposed to: for he has clearly no way of compelling those to reimburse him who have already paid him the fair price of his work." Mr. Buchanan has quoted, with great approbation, the following able passage from Mr. Malthus's work on population, which appears to me completely to answer his objection. "The price of labour, when left to find its natural level, is a most important political barometer, expressing the relation between the supply of provisions and the demand for them, between the quantity to be consumed and the number of consumers; and, taken on the average, independently of accidental circumstances, it further expresses, clearly, the wants of the society respecting population; that is, whatever may be the number of children to a marriage necessary to maintain exactly the present population, the price of labour will be just sufficient to support this number, or be above it, or below it, according to the state of the real funds for the maintenance of labour, whether stationary, progressive, or retrograde. Instead, however, of considering it in this light, we consider it as something which we may raise or depress at pleasure, something which depends principally on his majesty's justices of the peace. When an advance in the price of provisions already expresses that the demand is too great for the supply, in order to put the labourer in the same condition as before, we raise the price of labour, that is, we increase the demand, and are then much surprised that the price of provisions continues rising. In this, we act much in the same manner as if, when the quicksilver in the common weather-glass stood at *stormy*, we were to raise it by some forcible pressure to settled fair, and then be greatly astonished that it continued raining."

" The price of labour will express clearly the wants of the society respecting population; " it will be just sufficient to support the population, which at that time the state of the funds for the maintenance of labourers requires. If the labourer's wages were before only adequate to supply the requisite population, they will, after the tax, be inadequate to that supply, for he will not have the same funds to expend on his family. Labour will therefore rise, because the demand continues, and it is only by raising the price that the supply is not checked.

Nothing is more common than to see hats or malt rise when taxed; they rise because the requisite supply would not be afforded if they did not rise: so with labour, when wages are taxed, its price rises, because, if it did not, the requisite population would not be kept up. Does not Mr. Buchanan allow all that is contended for, when he says that " were he (the labourer) indeed reduced to a bare allowance of necessaries, he would then suffer no further abatement of his wages, as he could not on such conditions continue his race? " Suppose the circumstances of the country to be such that the lowest labourers are not only called upon to continue their race, but to increase it; their wages would be regulated accordingly. Can they multiply in the degree required if a tax takes from them a part of their wages, and reduces them to bare necessaries?

It is undoubtedly true that a taxed commodity will not rise in proportion to the tax if the demand for it diminish and if the quantity cannot be reduced. If metallic money were in general use, its value would not for a considerable time be increased by a tax, in proportion to the amount of the tax, because at a higher price the demand would be diminished and the quantity would not be diminished; and unquestionably the same cause frequently influences the wages of labour; the number of labourers cannot be rapidly increased or diminished in proportion to the increase or diminution of the fund which is to employ them; but in the case supposed, there is no necessary diminution of demand for labour, and if diminished, the demand does not abate in proportion to the tax. Mr. Buchanan forgets that the fund raised by the tax is employed by government in maintaining labourers, unproductive indeed, but still labourers. If labour were not to rise when wages are taxed, there would be a great increase in the competition for labour, because the owners of capital, who would have nothing to pay towards such a tax, would have the same funds for employing labour; whilst the government who received the tax would have an additional fund for the same purpose. Government and the people thus become competitors, and the consequence of their competition is a rise in the price of labour. The same number of men only will be employed, but they will be employed at additional wages.

If the tax had been laid at once on the people of capital, their fund for the maintenance of labour would have been diminished in the very same degree that the fund of government for that purpose had been increased; and therefore there would have been no rise in wages; for though there would be the same demand, there would not be the same competition. If when the tax were levied government at once exported the produce of it as a subsidy to a foreign state, and if therefore these funds were devoted to the maintenance of foreign and not of English labourers, such as soldiers, sailors, etc., etc.; then, indeed, there would be a diminished demand for labour, and wages might not increase although they were taxed; but the same thing would happen if the tax had been laid on consumable commodities, on the profits of stock, or if in any other manner the same sum had been raised to supply this subsidy: less labour could be employed at home. In one case wages are prevented from rising, in the other they must absolutely fall. But suppose the amount of a tax on wages were, after being raised on the labourers, paid gratuitously to their employers, it would increase their money fund for the maintenance of labour, but it would not increase either commodities or labour. It would consequently increase the competition amongst the employers of labour, and the tax would be ultimately attended with no loss either to master or labourer. The master would pay an increased price for labour; the addition which the labourer received would be paid as a tax to government, and would be again returned to the masters. It must, however, not be forgotten that the produce of taxes is generally wastefully expended, they are always obtained at the expense of the people's comforts and enjoyments, and commonly either diminish capital or retard its accumulation. By diminishing capital they tend to diminish the real fund destined for the maintenance of labour; and therefore to diminish the real demand for it. Taxes, then, generally, as far as they impair the real capital of the country, diminish the demand for labour, and therefore it is a probable, but not a necessary nor a peculiar consequence of a tax on wages, that though wages would rise, they would not rise by a sum precisely equal to the tax.

Adam Smith, as we have seen, has fully allowed that the effect of a tax on wages would be to raise wages by a sum at least equal to the tax, and would be finally, if not immediately, paid by the employer of labour. Thus far we fully agree; but we essentially differ in our views of the subsequent operation of such a tax.

" A direct tax upon the wages of labour, therefore," says Adam Smith, " though the labourer might perhaps pay it out of his hand, could not properly be said to be even advanced by him; at least if the demand for labour and the average price of provisions remained the same after the tax as before it. In all such cases, not only the tax but something more than the tax would in reality be advanced by the person who immediately employed him. The final payment would in different cases fall upon different persons. The rise which such a tax might occasion in the wages of manufacturing labour would be advanced by the master manufacturer, *who would be entitled and obliged to charge it with a profit upon the price of his goods.* The rise which such a tax might occasion in country labour would be advanced by the farmer, who, in order to maintain the same number of labourers as before, would be obliged to employ a greater capital. In order to get back this greater capital, *together with the ordinary profits of stock,* it would be necessary that he should retain a larger portion, or what comes to the same thing, the price of a larger portion, of the produce of the land, and consequently that he should pay less rent to the landlord. The final payment of this rise of wages would in this case fall upon the landlord, *together with the additional profits of the farmer who had advanced it.* In all cases, a direct tax upon the wages of labour must, in the long run, occasion both a greater reduction in the rent of land and a greater rise in the price of manufactured goods than would have followed from the proper assessment of a sum equal to the produce of the tax partly upon the rent of land and partly upon consumable commodities." Vol. iii. p. 337. In this passage it is asserted that the additional wages paid by farmers will ultimately fall on the landlords, who will receive a diminished rent; but that the additional wages paid by manufacturers will occasion a rise in the price of manufactured goods, and will therefore fall on the consumers of those commodities.

Now, suppose a society to consist of landlords, manufacturers, farmers, and labourers, the labourers, it is agreed, would be recompensed for the tax;—but by whom?—who would pay that portion which did not fall on the landlords?—the manufacturers could pay no part of it; for if the price of their commodities should rise in proportion to the additional wages they paid, they would be in a better situation after than before the tax. If the clothier, the hatter, the shoemaker, etc., should be each able to raise the price of their goods 10 per cent.—supposing 10 per cent. to recompense them completely for the additional wages they paid—if, as Adam Smith says, " they would be entitled and obliged to charge the additional wages *with a profit* upon the price of their goods," they could each consume as much as before of each other's goods, and therefore they would pay nothing towards the tax. If the clothier paid more for his hats and shoes, he would receive more for his cloth, and if the hatter paid more for his cloth and shoes, he would receive more for his hats. All manufactured commodities, then, would be bought by them with as much advantage as before, and inasmuch as corn would not be raised in price, which is Dr. Smith's supposition, whilst they had an additional sum to lay out upon its purchase, they would be benefited but not injured by such a tax.

If, then, neither the labourers nor the manufacturers would contribute towards such a tax; if the farmers would be also recompensed by a fall of rent, landlords alone must not only bear its whole weight, but they must also contribute to the increased gains of the manufacturers. To do this, however, they should consume all the manufactured commodities in the country, for the additional price charged on the whole mass is little more than the tax originally imposed on the labourers in manufactures.

Now, it will not be disputed that the clothier, the hatter, and all other manufacturers are consumers of each other's goods; it will not be disputed that labourers of all descriptions consume soap, cloth, shoes, candles, and various other commodities; it is therefore impossible that the whole weight of these taxes should fall on landlords only.

But if the labourers pay no part of the tax, and yet manufactured commodities rise in price, wages must rise, not only to compensate them for the tax, but for the increased price of manufactured necessaries, which, as far as it affects agricultural labour, will be a new cause for the fall of rent; and, as far as it affects manufacturing labour, for a further rise in the price of goods. This rise in the price of goods will again operate on wages, and the action and re-action, first of wages on goods, and then of goods on wages, will be extended without any assignable limits. The arguments by which this theory is supported lead to such absurd conclusions that it may at once be seen that the principle is wholly indefensible.

All the effects which are produced on the profits of stock and the wages of labour by a rise of rent and a rise of necessaries, in the natural progress of society and increasing difficulty of production, will equally follow from a rise of wages in consequence of taxation; and, therefore, the enjoyments of the labourer, as well as those of his employers, will be curtailed by the tax; and not by this tax particularly, but by every other which should raise an equal amount, as they would all tend to diminish the fund destined for the maintenance of labour.

The error of Adam Smith proceeds in the first place from supposing that all taxes paid by the farmer must necessarily fall on the landlord in the shape of a deduction from rent. On this subject I have explained myself most fully, and I trust that it has been shown, to the satisfaction of the reader, that since much capital is employed on the land which pays no rent, and since it is the result obtained by this capital which regulates the price of raw produce, no deduction can be made from rent; and, consequently, either no remuneration will be made to the farmer for a tax on wages, or if made, it must be made by an addition to the price of raw produce.

If taxes press unequally on the farmer, he will be enabled to raise the price of raw produce, to place himself on a level with those who carry on other trades; but a tax on wages, which would not affect him more than it would affect any other trade, could not be removed or compensated by a high price of raw produce; for the same reason which should induce him to raise the price of corn, namely, to remunerate himself for the tax, would induce the clothier to raise the price of cloth, the shoe-maker, hatter, and upholsterer to raise the price of shoes, hats, and furniture.

If they could all raise the price of their goods so as to remunerate themselves, with a profit, for the tax: as they are all consumers of each other's commodities, it is obvious that the tax could never be paid; for who would be the contributors if all were compensated?

I hope, then, that I have succeeded in showing that any tax which shall have the effect of raising wages will be paid by a diminution of profits, and, therefore, that a tax on wages is in fact a tax on profits.

This principle of the division of the produce of labour and capital between wages and profits, which I have attempted to establish, appears to me so certain, that excepting in the immediate effects, I should think it of little importance whether the profits of stock or the wages of labour, were taxed. By taxing the profits of stock you would probably alter the rate at which the funds for the maintenance of labour increase, and wages would be disproportioned to the state of that fund, by being too high. By taxing wages, the reward paid to the labourer would also be disproportioned to the state of that fund, by being too low. In the one case by a fall, and in the other by a rise in money wages, the natural equilibrium between profits and wages would be restored. A tax on wages, then, does not fall on the landlord, but it falls on the profits of stock: it does not "entitle and oblige the master manufacturer to charge it with a profit on the prices of his goods," for he will be unable to increase their price, and therefore he must himself wholly and without compensation pay such a tax.[1]

If the effect of taxes on wages be such as I have described, they do not merit the censure cast upon them by Dr. Smith. He observes of such taxes, " These, and some other taxes of the same kind, by raising the price of labour, are said to have ruined

---

[1] M. Say appears to have imbibed the general opinion on this subject. Speaking of corn, he says, " thence it results that its price influences the price of *all* other commodities. A farmer, a manufacturer, or a merchant employs a certain number of workmen who all have occasion to consume a certain quantity of corn. If the price of corn rises, he is obliged to raise, in an equal proportion, the price of his production."—Vol. i. p. 255.

the greater part of the manufactures of Holland. Similar taxes, though not quite so heavy, take place in the Milanese, in the states of Genoa, in the duchy of Modena, in the duchies of Parma, Placentia, and Guastalla, and in the ecclesiastical states. A French author of some note has proposed to reform the finances of his country by substituting in the room of other taxes this most ruinous of all taxes. ' There is nothing so absurd,' says Cicero, ' which has not sometimes been asserted by some philosophers.' " And in another place he says: " Taxes upon necessaries, by raising the wages of labour, necessarily tend to raise the price of all manufactures, and consequently to diminish the extent of their sale and consumption." They would not merit this censure, even if Dr. Smith's principle were correct, that such taxes would enhance the prices of manufactured commodities; for such an effect could be only temporary, and would subject us to no disadvantage in our foreign trade. If any cause should raise the price of a few manufactured commodities, it would prevent or check their exportation; but if the same cause operated generally on all, the effect would be merely nominal, and would neither interfere with their relative value, nor in any degree diminish the stimulus to a trade of barter, which all commerce, both foreign and domestic, really is.

I have already attempted to show that when any cause raises the prices of all commodities the effects are nearly similar to a fall in the value of money. If money falls in value all commodities rise in price; and if the effect is confined to one country, it will affect its foreign commerce in the same way as a high price of commodities caused by general taxation; and, therefore, in examining the effects of a low value of money confined to one country, we are also examining the effects of a high price of commodities confined to one country. Indeed, Adam Smith was fully aware of the resemblance between these two cases, and consistently maintained that the low value of money, or, as he calls it, of silver in Spain, in consequence of the prohibition against its exportation, was very highly prejudicial to the manufactures and foreign commerce of Spain. " But that degradation in the value of silver, which being the effect either of the peculiar situation, or of the political institutions of a particular country, takes place only in that country, is a matter of very great consequence, which, far from tending to make anybody really richer, tends to make everybody really poorer. *The rise in the money price of all commodities, which is in this case peculiar to that country*, tends to discourage more or less every sort of industry which is carried on within it, and to enable foreign nations, by furnishing almost all sorts of goods for a smaller quantity of silver than its own workmen can afford to do, to undersell them not only in the foreign but even in the home market." Vol. ii. p. 278.

One, and I think the only one, of the disadvantages of a low value of silver in a country, proceeding from a forced abundance, has been ably explained by Dr. Smith. If the trade in gold and silver were free, " the gold and silver which would go abroad would not go abroad for nothing, but would bring back an equal value of goods of some kind or another. Those goods, too, would not be all matters of mere luxury and expense to be consumed by idle people, who produce nothing in return for their consumption. As the real wealth and revenue of idle people would not be augmented by this extraordinary exportation of gold and silver, so would neither their consumption be augmented by it. Those goods would—probably the greater part of them, and certainly some part of them—consist in materials, tools, and provisions, for the employment and maintenance of industrious people, who would reproduce with a profit the full value of their consumption. A part of the dead stock of the society would thus be turned into active stock, and would put into motion a greater quantity of industry than had been employed before."

By not allowing a free trade in the precious metals when the prices of commodities are raised, either by taxation, or by the influx of the precious metals, you prevent a part of the dead stock of the society from being turned into active stock—you prevent a greater quantity of industry from being employed. But this is the whole amount of the evil—an evil never felt by those countries where the exportation of silver is either allowed or connived at.

The exchanges between countries are at par only whilst they have precisely that quantity of currency which, in the actual situation of things, they should have to carry on the circulation of their commodities. If the trade in the precious metals were perfectly free, and money could be exported without any expense whatever, the exchanges could be no otherwise in every country than at par. If the trade in the precious metals were perfectly free—if they were generally used in circulation, even with the expenses of transporting them, the exchange could never in any of them deviate more from par than by these expenses. These principles, I believe, are now nowhere disputed. If a country used paper money not exchangeable for specie, and, therefore, not regulated by any fixed standard, the exchanges in that country might deviate from par in the same proportion as its money might be multiplied beyond that quantity which would have been allotted to it by general commerce, if the trade in money had been free, and the precious metals had been used, either for money, or for the standard of money.

If by the general operations of commerce, 10 millions of pounds sterling, of a known weight and fineness of bullion, should be the portion of England, and 10 millions of paper pounds were substituted, no effect would be produced on the exchange; but if by the abuse of the power of issuing paper money, 11 millions of pounds should be employed in the circulation, the exchange would be 9 per cent. against England; if 12 millions were employed, the exchange would be 16 per cent.; and if 20 millions, the exchange would be 50 per cent. against England. To produce this effect it is not, however, necessary that paper money should be employed: any cause which retains in circulation a greater quantity of pounds than would have circulated if commerce had been free, and the precious metals of a known weight and fineness had been used, either for money or for the standard of money, would exactly produce the same effects. Suppose that by clipping the money each pound did not contain the quantity of gold or silver which by law it should contain, a greater number of such pounds might be employed in the circulation than if they were not clipped. If from each pound one-tenth were taken away, 11 millions of such pounds might be used instead of 10; if two-tenths were taken away, 12 millions might be employed; and if one-half were taken away, 20 millions might not be found superfluous. If the latter sum were used instead of 10 millions, every commodity in England would be raised to double its former price, and the exchange would be 50 per cent. against England; but this would occasion no disturbance in foreign commerce, nor discourage the manufacture of any one commodity. If, for example, cloth rose in England from £20 to £40 per piece, we should just as freely export it after as before the rise, for a compensation of 50 per cent. would be made to the foreign purchaser in the exchange; so that with £20 of his money, he could purchase a bill which would enable him to pay a debt of £40 in England. In the same manner, if he exported a commodity which cost £20 at home, and which sold in England for £40, he would only receive £20, for £40 in England would only purchase a bill for £20 on a foreign country. The same effects would follow from whatever cause 20 millions could be forced to perform the business of circulation in England if 10 millions only were necessary. If so absurd a law as the prohibition of the exportation of the precious metals could be enforced, and the consequence of such prohibition were to force 11 millions of good pounds, fresh from the mint, instead of 10, into circulation, the exchange would be 9 per cent. against England; if 12 millions, 16 per cent.; and if 20 millions, 50 per cent. against England. But no discouragement would be given to the manufactures of England; if home commodities sold at a high price in England, so would foreign commodities; and whether they were high or low would be of little importance to the foreign exporter and importer, whilst he would, on the one hand, be obliged to allow a compensation in the exchange when his commodities sold at a dear rate, and would receive the same compensation when he was obliged to purchase English commodities at a high price. The sole disadvantage, then, which could happen to a country from retaining, by prohibitory laws, a greater quantity of gold and silver in circulation than would otherwise remain there, would be the loss which it would sustain from employing a portion of its capital unproductively instead of employing it productively. In the form of money, this capital is productive of no profit; in the form of materials, machinery, and food, for which it might be exchanged, it would be productive of revenue, and would add to the wealth and the resources of the state. Thus, then, I hope, I have satisfactorily proved that a comparatively low price of the precious metals, in consequence of taxation, or in other words, a generally high price of commodities, would be of no disadvantage to a state, as a part of the metals would be exported, which, by raising their value, would again lower the prices of commodities. And further, that if they were not exported, if by prohibitory laws they could be retained in a country, the effect on the exchange would counterbalance the effect of high prices. If, then, taxes on necessaries and on wages would not raise the prices of all commodities on which labour was expended, they cannot be condemned on such grounds; and moreover, even if the opinion given by Adam Smith, that they would have such an effect, were well founded, they would be in no degree injurious on that account. They would be objectionable for no other reason than those which might be justly urged against taxes of any other description.

The landlords, as such, would be exempted from the burden of the tax; but as far as they directly employed labour in the expenditure of their revenues, by supporting gardeners, menial servants, etc., they would be subject to its operation.

It is undoubtedly true that "taxes upon luxuries have no tendency to raise the price of any other commodities, except that of the commodities taxed "; but it is not true " that taxes upon necessaries, by raising the wages of labour, necessarily tend to raise the price of all manufactures." It is true that " taxes upon luxuries are finally paid by the consumers of the commodities taxed, without any retribution. They fall indifferently upon every species of revenue, the wages of labour, the profits of stock, and the rent of land "; but it is not true " that taxes upon necessaries, *so far as they affect the labouring poor,* are finally paid partly by landlords in the diminished rent of their lands, and partly by rich consumers, whether landlords or others, in the advanced price of manufactured goods "; for, *so far as these taxes affect the labouring poor,* they will be almost wholly paid by the diminished profits of stock, a small part only being paid by the labourers themselves in the diminished demand for labour, which taxation of every kind has a tendency to produce.

It is from Dr. Smith's erroneous view of the effect of those taxes that he has been led to the conclusion that " the middling and superior ranks of people, if they understood their own interest, ought always to oppose all taxes upon the necessaries of life, as well as all direct taxes upon the wages of labour." This conclusion follows from his reasoning, " that the final payment of both one and the other falls altogether upon themselves, and always with a considerable overcharge. They fall heaviest upon the landlords,[1] who always pay in a double capacity; in that of landlords by the reduction of their rent, and in that of rich consumers by the increase of their expense. The observation of Sir Matthew Decker, that certain taxes are, in the price of certain goods, sometimes repeated and accumulated four or five times, is perfectly just with regard to taxes upon the necessaries of life. In the price of leather, for example, you must pay, not only for the tax upon the leather of your own shoes, but for a part of that upon those of the shoemaker and the tanner. You must pay, too, for the tax upon the salt, upon the soap, and upon the candles which those workmen consume while employed in your service, and for the tax upon the leather

---

[1] So far from this being true, they would scarcely affect the landlords and stockholder.

which the salt-maker, the soap-maker, and the candle-maker consume while employed in their service."

Now as Dr. Smith does not contend that the tanner, the salt-maker, the soap-maker, and the candle-maker will either of them be benefited by the tax on leather, salt, soap, and candles; and as it is certain that government will receive no more than the tax imposed, it is impossible to conceive that more can be paid by the public upon whomsoever the tax may fall. The rich consumers may, and indeed will, pay for the poor consumer, but they will pay no more than the whole amount of the tax; and it is not in the nature of things that " the tax should be repeated and accumulated four or five times."

A system of taxation may be defective; more may be raised from the people than what finds its way into the coffers of the state, as a part, in consequence of its effect on prices, may possibly be received by those who are benefited by the peculiar mode in which taxes are laid. Such taxes are pernicious, and should not be encouraged; for it may be laid down as a principle, that when taxes operate justly, they conform to the first of Dr. Smith's maxims, and raise from the people as little as possible beyond what enters into the public treasury of the state. M. Say says, " others offer plans of finance, and propose means for filling the coffers of the sovereign, without any charge to his subjects. But unless a plan of finance is of the nature of a commercial undertaking, it cannot give to government more than it takes away either from individuals or from government itself, under some other form. Something cannot be made out of nothing by the stroke of a wand. In whatever way an operation may be disguised, whatever forms we may constrain a value to take, whatever metamorphosis we may make it undergo, we can only have a value by creating it, or by taking it from others. The very best of all plans of finance is to spend little, and the best of all taxes is that which is the least in amount."

Dr. Smith uniformly, and I think justly, contends that the labouring classes cannot materially contribute to the burdens of the state. A tax on necessaries, or on wages, will therefore be shifted from the poor to the rich: if then the meaning of Dr. Smith is, " that certain taxes are in the price of certain goods sometimes repeated, and accumulated four or five times," for the purpose only of accomplishing this end, namely, the transference of the tax from the poor to the rich, they cannot be liable to censure on that account.

Suppose the just share of the taxes of a rich consumer to be

£100, and that he would pay it directly if the tax were laid on income, on wine, or on any other luxury, he would suffer no injury if, by the taxation of necessaries, he should be only called upon for the payment of £25, as far as his own consumption of necessaries and that of his family was concerned; but should be required to repeat this tax three times, by paying an additional price for other commodities to remunerate the labourers, or their employers, for the tax which they have been called upon to advance. Even in that case the reasoning is inconclusive: for if there be no more paid than what is required by government, of what importance can it be to the rich consumer whether he pay the tax directly, by paying an increased price for an object of luxury, or indirectly, by paying an increased price for the necessaries and other commodities he consumes? If more be not paid by the people than what is received by government, the rich consumer will only pay his equitable share; if more is paid, Adam Smith should have stated by whom it is received; but his whole argument is founded in error, for the prices of commodities would not be raised by such taxes.

M. Say does not appear to me to have consistently adhered to the obvious principle which I have quoted from his able work; for in the next page, speaking of taxation, he says, "When it is pushed too far, it produces this lamentable effect, it deprives the contributor of a portion of his riches, without enriching the state. This is what we may comprehend if we consider that every man's power of consuming, whether productively or not, is limited by his income. He cannot then be deprived of a part of his income without being obliged proportionally to reduce his consumption. Hence arises a diminution of demand for those goods which he no longer consumes, and particularly for those on which the tax is imposed. From this diminution of demand there results a diminution of production, and consequently of taxable commodities. The contributor then will lose a portion of his enjoyments; the producer a portion of his profits; and the treasury a portion of its receipts."

M. Say instances the tax on salt in France previous to the revolution; which, he says, diminished the production of salt by one half. If, however, less salt was consumed, less capital was employed in producing it; and, therefore, though the producer would obtain less profit on the production of salt, he would obtain more on the production of other things. If a tax, however burdensome it may be, falls on revenue, and not on capital, it does not diminish demand, it only alters the nature of it. It enables government to consume as much of the produce of the land and labour of the country as was before consumed by the individuals who contribute to the tax, an evil sufficiently great without overcharging it. If my income is £1000 per annum, and I am called upon for £100 per annum for a tax, I shall only be able to demand nine-tenths of the quantity of goods which I before consumed, but I enable government to demand the other tenth. If the commodity taxed be corn, it is not necessary that my demand for corn should diminish, as I may prefer to pay £100 per annum more for my corn, and to the same amount abate in my demand for wine, furniture, or any other luxury.[1] Less capital will consequently be employed in the wine or upholstery trade, but more will be employed in manufacturing those commodities, on which the taxes levied by government will be expended.

M. Say says that M. Turgot, by reducing the market dues on fish (*les droits d'entrée et de halle sur la marée*) in Paris one half, did not diminish the amount of their produce, and that consequently the consumption of fish must have doubled. He infers from this that the profits of the fishermen and those engaged in the trade must also have doubled, and that the income of the country must have increased by the whole amount of these increased profits; and by giving a stimulus to accumulation, must have increased the resources of the state.[2]

Without calling in question the policy which dictated this alteration of the tax, I have my doubts whether it gave any great stimulus to accumulation. If the profits of the fishermen and others engaged in the trade were doubled in consequence of more fish being consumed, capital and labour must have been withdrawn from other occupations to engage them in this particular trade. But in those occupations capital and labour

[1] M. Say says, " that the tax added to the price of a commodity raises its price. Every increase in the price of a commodity necessarily reduces the number of those who are able to purchase it, or at least the quantity they will consume of it." This is by no means a necessary consequence. I do not believe that if bread were taxed the consumption of bread would be diminished, more than if cloth, wine, or soap were taxed.

[2] The following remark of the same author appears to me equally erroneous: " When a high duty is laid on cotton the production of all those goods of which cotton is the basis is diminished. If the total value added to cotton in its various manufactures, in a particular country, amounted to 100 millions of francs per annum, and the effect of the tax was to diminish the consumption one half, then the tax would deprive that country every year of 50 million of francs, in addition to the sum received by government."—Vol. ii. p. 314.

were productive of profits, which must have been given up when they were withdrawn. The ability of the country to accumulate was only increased by the difference between the profits obtained in the business in which the capital was newly engaged, and those obtained in that from which it was withdrawn.

Whether taxes be taken from revenue or capital they diminish the taxable commodities of the state. If I cease to expend £100 on wine, because by paying a tax of that amount I have enabled government to expend £100 instead of expending it myself, one hundred pounds' worth of goods are necessarily withdrawn from the list of taxable commodities. If the revenue of the individuals of a country be 10 millions, they will have at least 10 millions' worth of taxable commodities. If, by taxing some, one million be transferred to the disposal of government, their revenue will still be nominally 10 millions, but they will remain with only nine millions' worth of taxable commodities. There are no circumstances under which taxation does not abridge the enjoyments of those on whom the taxes ultimately fall, and no means by which those enjoyments can again be extended but the accumulation of new revenue.

Taxation can never be so equally applied as to operate in the same proportion on the value of all commodities, and still to preserve them at the same relative value. It frequently operates very differently from the intention of the legislature by its indirect effects. We have already seen that the effect of a direct tax on corn and raw produce is, if money be also produced in the country, to raise the price of all commodities in proportion as raw produce enters into their composition, and thereby to destroy the natural relation which previously existed between them. Another indirect effect is that it raises wages and lowers the rate of profits; and we have also seen, in another part of this work, that the effect of a rise of wages and a fall of profits is to lower the money prices of those commodities which are produced in a greater degree by the employment of fixed capital.

That a commodity, when taxed, can no longer be so profitably exported is so well understood that a drawback is frequently allowed on its exportation, and a duty laid on its importation. If these drawbacks and duties be accurately laid, not only on the commodities themselves, but on all which they indirectly affect, then, indeed, there will be no disturbance in the value of the precious metals. Since we could as readily export a commodity after being taxed as before, and since no peculiar facility would be given to importation, the precious metals would not, more than before, enter into the list of exportable commodities.

Of all commodities none are perhaps so proper for taxation as those which, either by the aid of nature or art, are produced with peculiar facility. With respect to foreign countries, such commodities may be classed under the head of those which are not regulated in their price by the quantity of labour bestowed, but rather by the caprice, the tastes, and the power of the purchasers. If England had more productive tin mines than other countries, or if, from superior machinery or fuel, she had peculiar facilities in manufacturing cotton goods, the prices of tin and cotton goods would still in England be regulated by the comparative quantity of labour and capital required to produce them, and the competition of our merchants would make them very little dearer to the foreign consumer. Our advantage in the production of these commodities might be so decided that probably they could bear a very great additional price in the foreign market without very materially diminishing their consumption. This price they never could attain, whilst competition was free at home, by any other means but by a tax on their exportation. This tax would fall wholly on foreign consumers, and part of the expenses of the government of England would be defrayed by a tax on the land and labour of other countries. The tax on tea, which at present is paid by the people of England, and goes to aid the expenses of the government of England, might, if laid in China on the exportation of the tea, be diverted to the payment of the expenses of the government of China.

Taxes on luxuries have some advantage over taxes on necessaries. They are generally paid from income, and therefore do not diminish the productive capital of the country. If wine were much raised in price in consequence of taxation, it is probable that a man would rather forego the enjoyments of wine than make any important encroachments on his capital to be enabled to purchase it. They are so identified with price that the contributor is hardly aware that he is paying a tax. But they have also their disadvantages. First, they never reach capital, and on some extraordinary occasions it may be expedient that even capital should contribute towards the public exigencies; and, secondly, there is no certainty as to the amount of the tax, for it may not reach even income. A man intent on saving will exempt himself from a tax on wine by giving up the use of it. The income of the country may be undiminished, and yet the state may be unable to raise a shilling by the tax.

Whatever habit has rendered delightful will be relinquished with reluctance, and will continue to be consumed notwith-standing a very heavy tax; but this reluctance has its limits, and experience every day demonstrates that an increase in the nominal amount of taxation often diminishes the produce. One man will continue to drink the same quantity of wine, though the price of every bottle should be raised three shillings, who would yet relinquish the use of wine rather than pay four. Another will be content to pay four, yet refuse to pay five shillings. The same may be said of other taxes on luxuries: many would pay a tax of £5 for the enjoyment which a horse affords, who would not pay £10 or £20. It is not because they cannot pay more that they give up the use of wine and of horses, but because they will not pay more. Every man has some standard in his own mind by which he estimates the value of his enjoyments, but that standard is as various as the human character. A country whose financial situation has become extremely artificial, by the mischievous policy of accumulating a large national debt, and a consequently enormous taxation, is particularly exposed to the inconvenience attendant on this mode of raising taxes. After visiting with a tax the whole round of luxuries; after laying horses, carriages, wine, servants, and all the other enjoyments of the rich under contribution; a minister is induced to have recourse to more direct taxes, such as income and property taxes, neglecting the golden maxim of M. Say, " that the very best of all plans of finance is to spend little, and the best of all taxes is that which is the least in amount."

CHAPTER XVII

TAXES ON OTHER COMMODITIES THAN RAW PRODUCE

On the same principle that a tax on corn would raise the price of corn, a tax on any other commodity would raise the price of that commodity. If the commodity did not rise by a sum equal to the tax, it would not give the same profit to the producer which he had before, and he would remove his capital to some other employment.

The taxing of all commodities, whether they be necessaries or luxuries, will, while money remains at an unaltered value, raise their prices by a sum at least equal to the tax.[1] A tax on the manufactured necessaries of the labourer would have the same effect on wages as a tax on corn, which differs from other necessaries only by being the first and most important on the list; and it would produce precisely the same effects on the profits of stock and foreign trade. But a tax on luxuries would have no other effect than to raise their price. It would fall wholly on the consumer, and could neither increase wages nor lower profits.

Taxes which are levied on a country for the purpose of supporting war, or for the ordinary expenses of the state, and which are chiefly devoted to the support of unproductive labourers, are taken from the productive industry of the country; and every saving which can be made from such expenses will be

[1] It is observed by M. Say, " that a manufacturer is not enabled to make the consumer pay the whole tax levied on his commodity, because its increased price will diminish its consumption." Should this be the case, should the consumption be diminished, will not the supply also speedily be diminished? Why should the manufacturer continue in the trade if his profits are below the general level? M. Say appears here also to have forgotten the doctrine which he elsewhere supports, " that the cost of production determines the price, below which commodities cannot fall for any length of time, because production would be then either suspended or diminished."—Vol. ii. p. 26.

" The tax in this case falls then partly on the consumer, who is obliged to give more for the commodity taxed, and partly on the producer, who, after deducting the tax, will receive less. The public treasury will be benefited by what the purchaser pays in addition, and also by the sacrifice which the producer is obliged to make of a part of his profits. It is the effort of gunpowder, which acts at the same time on the bullet which it projects and on the gun which it causes to recoil."—Vol. ii. p. 333.

generally added to the income, if not to the capital of the contributors. When, for the expenses of a year's war, twenty millions are raised by means of a loan, it is the twenty millions which are withdrawn from the productive capital of the nation. The million per annum which is raised by taxes to pay the interest of this loan is merely transferred from those who pay it to those who receive it, from the contributor to the tax to the national creditor. The real expense is the twenty millions, and not the interest which must be paid for it.[1] Whether the interest be or be not paid, the country will neither be richer nor poorer. Government might at once have required the twenty millions in the shape of taxes; in which case it would not have been necessary to raise annual taxes to the amount of a million. This, however, would not have changed the nature of the transaction. An individual, instead of being called upon to pay £100 per annum, might have been obliged to pay £2000 once for all. It might also have suited his convenience rather to borrow this £2000, and to pay £100 per annum for interest to the lender, than to spare the larger sum from his own funds. In one case, it is a private transaction between A and B, in the other government guarantees to B the payment of interest to be equally paid by A. If the transaction had been of a private nature, no public record would be kept of it, and it would be a matter of comparative indifference to the country whether A faithfully performed his contract to B or unjustly retained the £100 per annum in his own possession. The country would have a general interest in the faithful performance of a contract, but with respect to the national wealth it would have no other interest than whether A or B would make this £100 most productive; but on this question it would neither have the right

[1] " Melon says that the debts of a nation are debts due from the right hand to the left, by which the body is not weakened. It is true that the general wealth is not diminished by the payment of the interest on arrears of the debt: The dividends are a value which passes from the hand of the contributor to the national creditor: Whether it be the national creditor or the contributor who accumulates or consumes it is, I agree, of little importance to the society; but the principal of the debt—what has become of that? It exists no more. The consumption which has followed the loan has annihilated a capital which will never yield any further revenue. The society is deprived not of the amount of interest, since that passes from one hand to the other, but of the revenue from a destroyed capital. This capital, if it had been employed productively by him who lent it to the state, would equally have yielded him an income, but that income would have been derived from a real production, and would not have been furnished from the pocket of a fellow citizen."—Say, vol. ii. p. 357. This is both conceived and expressed in the true spirit of the science.

nor the ability to decide. It might be possible that, if A retained it for his own use, he might squander it unprofitably, and if it were paid to B he might add it to his capital and employ it productively. And the converse would also be possible; B might squander it, and A might employ it productively. With a view to wealth only, it might be equally or more desirable that A should or should not pay it; but the claims of justice and good faith, a greater utility, are not to be compelled to yield to those of a less; and accordingly, if the state were called upon to interfere, the courts of justice would oblige A to perform his contract. A debt guaranteed by the nation differs in no respect from the above transaction. Justice and good faith demand that the interest of the national debt should continue to be paid, and that those who have advanced their capitals for the general benefit should not be required to forego their equitable claims on the plea of expediency.

But independently of this consideration, it is by no means certain that political utility would gain anything by the sacrifice of political integrity; it does by no means follow that the party exonerated from the payment of the interest of the national debt would employ it more productively than those to whom indisputably it is due. By cancelling the national debt, one man's income might be raised from £1000 to £1500, but another man's would be lowered from £1500 to £1000. These two men's incomes now amount to £2500; they would amount to no more then. If it be the object of government to raise taxes, there would be precisely the same taxable capital and income in one case as in the other. It is not, then, by the payment of the interest on the national debt that a country is distressed, nor is it by the exoneration from payment that it can be relieved. It is only by saving from income, and retrenching in expenditure, that the national capital can be increased; and neither the income would be increased nor the expenditure diminished by the annihilation of the national debt. It is by the profuse expenditure of government and of individuals, and by loans, that the country is impoverished; every measure, therefore, which is calculated to promote public and private economy will relieve the public distress; but it is error and delusion to suppose that a real national difficulty can be removed by shifting it from the shoulders of one class of the community, who justly ought to bear it, to the shoulders of another class, who, upon every principle of equity, ought to bear no more than their share.

From what I have said, it must not be inferred that I consider the system of borrowing as the best calculated to defray the extraordinary expenses of the state. It is a system which tends to make us less thrifty—to blind us to our real situation. If the expenses of a war be 40 millions per annum, and the share which a man would have to contribute towards that annual expense were £100, he would endeavour, on being at once called upon for his portion, to save speedily the £100 from his income. By the system of loans, he is called upon to pay only the interest of this £100, or £5 per annum, and considers that he does enough by saving this £5 from his expenditure, and then deludes himself with the belief that he is as rich as before. The whole nation, by reasoning and acting in this manner, save only the interest of 40 millions, or two millions; and thus not only lose all the interest or profit which 40 millions of capital, employed productively, would afford, but also 38 millions, the difference between their savings and expenditure. If, as I before observed, each man had to make his own loan, and contribute his full proportion to the exigencies of the state, as soon as the war ceased taxation would cease, and we should immediately fall into a natural state of prices. Out of his private funds, A might have to pay to B interest for the money he borrowed of him during the war to enable him to pay his quota of the expense; but with this the nation would have no concern.

A country which has accumulated a large debt is placed in a most artificial situation; and although the amount of taxes, and the increased price of labour, may not, and I believe does not, place it under any other disadvantage with respect to foreign countries, except the unavoidable one of paying those taxes, yet it becomes the interest of every contributor to withdraw his shoulder from the burthen, and to shift this payment from himself to another; and the temptation to remove himself and his capital to another country, where he will be exempted from such burthens, becomes at last irresistible, and overcomes the natural reluctance which every man feels to quit the place of his birth and the scene of his early associations. A country which has involved itself in the difficulties attending this artificial system would act wisely by ransoming itself from them at the sacrifice of any portion of its property which might be necessary to redeem its debt. That which is wise in an individual is wise also in a nation. A man who has £10,000, paying him an income of £500, out of which he has to pay £100 per annum towards the interest of the debt, is really worth only £8000, and would be equally rich, whether he continued to pay

£100 per annum, or at once, and for only once, sacrificed £2000. But where, it is asked, would be the purchaser of the property which he must sell to obtain this £2000? The answer is plain: the national creditor, who is to receive this £2000, will want an investment for his money, and will be disposed either to lend it to the landholder, or manufacturer, or to purchase from them a part of the property of which they have to dispose. To such a payment the stockholders themselves would largely contribute. This scheme has been often recommended, but we have, I fear, neither wisdom enough, nor virtue enough, to adopt it. It must, however, be admitted, that during peace, our unceasing efforts should be directed towards paying off that part of the debt which has been contracted during war; and that no temptation of relief, no desire of escape from present, and I hope temporary, distresses should induce us to relax in our attention to that great object.

No sinking fund can be efficient for the purpose of diminishing the debt if it be not derived from the excess of the public revenue over the public expenditure. It is to be regretted that the sinking fund in this country is only such in name; for there is no excess of revenue above expenditure. It ought, by economy, to be made what it is professed to be, a really efficient fund for the payment of the debt. If, on the breaking out of any future war, we shall not have very considerably reduced our debt, one of two things must happen, either the whole expenses of that war must be defrayed by taxes raised from year to year, or we must, at the end of that war, if not before, submit to a national bankruptcy; not that we shall be unable to bear any large additions to the debt; it would be difficult to set limits to the powers of a great nation; but assuredly there are limits to the price, which in the form of perpetual taxation, individuals will submit to pay for the privilege merely of living in their native country.[1]

When a commodity is at a monopoly price it is at the very highest price at which the consumers are willing to purchase it.

---

[1] " Credit, in general, is good, as it allows capitals to leave those hands where they are not usefully employed, to pass into those where they will be made productive: it diverts a capital from an employment useful only to the capitalist, such as an investment in the public funds, to make it productive in the hands of industry. It facilitates the employments of all capitals, and leaves none unemployed."—*Economie Politique*, p. 463, vol. ii. 4th edition.—This must be an oversight of M. Say. The capital of the stockholder can never be made productive—it is, in fact, no capital, If he were to sell his stock, and employ the capital he obtained for it, productively, he could only do so by detaching the capital of the buyer of his stock from a productive employment.

Commodities are only at a monopoly price when by no possible device their quantity can be augmented; and when, therefore, the competition is wholly on one side — amongst the buyers. The monopoly price of one period may be much lower or higher than the monopoly price of another, because the competition amongst the purchasers must depend on their wealth, and their tastes and caprices. Those peculiar wines which are produced in very limited quantity, and those works of art which, from their excellence or rarity, have acquired a fanciful value, will be exchanged for a very different quantity of the produce of ordinary labour, according as the society is rich or poor, as it possesses an abundance or scarcity of such produce, or as it may be in a rude or polished state. The exchangeable value therefore of a commodity which is at a monopoly price is nowhere regulated by the cost of production.

Raw produce is not at a monopoly price, because the market price of barley and wheat is as much regulated by their cost of production as the market price of cloth and linen. The only difference is this, that one portion of the capital employed in agriculture regulates the price of corn, namely, that portion which pays no rent; whereas, in the production of manufactured commodities, every portion of capital is employed with the same results; and as no portion pays rent, every portion is equally a regulator of price: corn, and other raw produce, can be augmented, too, in quantity, by the employment of more capital on the land, and therefore they are not at a monopoly price. There is competition among the sellers, as well as amongst the buyers. This is not the case in the production of those rare wines, and those valuable specimens of art, of which we have been speaking; their quantity cannot be increased, and their price is limited only by the extent of the power and will of the purchasers. The rent of these vineyards may be raised beyond any moderately assignable limits, because no other land being able to produce such wines, none can be brought into competition with them.

The corn and raw produce of a country may, indeed, for a time, sell at a monopoly price; but they can do so permanently only when no more capital can be profitably employed on the lands, and when, therefore, their produce cannot be increased. At such time, every portion of land in cultivation, and every portion of capital employed on the land, will yield a rent, differing, indeed, in proportion to the difference in the return. At such a time, too, any tax which may be imposed on the farmer will fall on rent, and not on the consumer. He cannot raise the price of his corn, because, by the supposition, it is already at the highest price at which the purchasers will or can buy it. He will not be satisfied with a lower rate of profits than that obtained by other capitalists, and, therefore, his only alternative will be to obtain a reduction of rent or to quit his employment.

Mr. Buchanan considers corn and raw produce as at a monopoly price, because they yield a rent: all commodities which yield a rent, he supposes, must be at a monopoly price; and thence he infers that all taxes on raw produce would fall on the landlord, and not on the consumer. "The price of corn," he says, "which always affords a rent, being in no respect influenced by the expenses of its production, those expenses must be paid out of the rent; and when they rise or fall, therefore, the consequence is not a higher or lower price, but a higher or a lower rent. In this view, all taxes on farm servants, horses, or the implements of agriculture are in reality land-taxes—the burden falling on the farmer during the currency of his lease, and on the landlord when the lease comes to be renewed. In like manner, all those improved implements of husbandry which save expense to the farmer, such as machines for thrashing and reaping, whatever gives him easier access to the market, such as good roads, canals, and bridges, though they lessen the original cost of corn, do not lessen its market price. Whatever is saved by those improvements, therefore, belongs to the landlord as part of his rent.

It is evident that if we yield to Mr. Buchanan the basis on which his argument is built, namely, that the price of corn always yields a rent, all the consequences which he contends for would follow of course. Taxes on the farmer would then fall, not on the consumer, but on rent; and all improvements in husbandry would increase rent: but I hope I have made it sufficiently clear that, until a country is cultivated in every part, and up to the highest degree, there is always a portion of capital employed on the land which yields no rent, and that it is this portion of capital, the result of which, as in manufactures, is divided between profits and wages, that regulates the price of corn. The price of corn, then, which does not afford a rent, being influenced by the expenses of its production, those expenses cannot be paid out of rent. The consequence, therefore, of those expenses increasing, is a higher price, and not a lower rent.[1]

[1] " Manufacturing industry increases its produce in proportion to the demand, and the price falls; *but the produce of land cannot be so increased ;*

It is remarkable that both Adam Smith and Mr. Buchanan,
who entirely agree that taxes on raw produce, a land-tax, and
tithes, all fall on the rent of land, and not on the consumers of
raw produce, should nevertheless admit that taxes on malt
would fall on the consumer of beer, and not on the rent of the
landlord. Adam Smith's argument is so able a statement of
the view which I take of the subject of the tax on malt, and
every other tax on raw produce, that I cannot refrain from
offering it to the attention of the reader.

" The rent and profits of barley land must always be nearly
equal to those of other equally fertile and equally well cultivated
land. If they were less, some part of the barley land would
soon be turned to some other purpose; and if they were greater,
more land would soon be turned to the raising of barley. When
the ordinary price of any particular produce of land is at what
may be called a monopoly price, a tax upon it necessarily reduces
the rent and profit [1] of the land which grows it. A tax upon
the produce of those precious vineyards, of which the wine falls
so much short of the effectual demand that its price is always
above the natural proportion to that of other equally fertile and
equally well cultivated land, would necessarily reduce the rent
and profit [1] of those vineyards. The price of the wines being
already the highest that could be got for the quantity commonly
sent to market, it could not be raised higher without diminishing
that quantity; and the quantity could not be diminished without still greater loss, because the lands could not be turned to
any other equally valuable produce. The whole weight of the
tax, therefore, would fall upon the rent and profit; [1] properly
upon the *rent* of the vineyard."—" But the ordinary price of
barley has never been a monopoly price; and the rent and profits
of barley land have never been above their natural proportion
to those of other equally fertile and equally well cultivated land.
The different taxes which have been imposed upon malt, beer,
and ale *have never lowered the price of barley;* have never
reduced the rent and profit [1] of barley land. The price of malt

and a high price is still necessary to prevent the consumption from exceeding the supply."—Buchanan, vol. iv. p. 40. Is it possible that Mr.
Buchanan can seriously assert that the produce of the land cannot be
increased if the demand increases?

[1] I wish the word " profit " had been omitted. Dr. Smith must suppose
the profits of the tenants of these precious vineyards to be above the
general rate of profits. If they were not, they would not pay the tax,
unless they could shift it either to the landlord or consumer.

to the brewer has constantly risen in proportion to the taxes imposed upon it; and those taxes, together with the different duties upon beer and ale, have constantly either raised the price, or, what comes to the same thing, reduced the quality of those commodities to the consumer. The final payment of those taxes has fallen constantly upon the consumer and not upon the producer." On this passage Mr. Buchanan remarks, " A duty on malt never could reduce the price of barley, because, unless as much could be made of barley by malting it as by selling it unmalted, the quantity required would not be brought to market. It is clear, therefore, that the price of malt must rise in proportion to the tax imposed on it, as the demand could not otherwise be supplied. The price of barley, however, is just as much a monopoly price as that of sugar; they both yield a rent, and the market price of both has equally lost all connection with the original cost."

It appears, then, to be the opinion of Mr. Buchanan, that a tax on malt would raise the price of malt, but that a tax on the barley from which malt is made would not raise the price of barley; and, therefore, if malt is taxed, the tax will be paid by the consumer; if barley is taxed, it will be paid by the land-lord, as he will receive a diminished rent. According to Mr. Buchanan, then, barley is at a monopoly price at the highest price which the purchasers are willing to give for it; but malt made of barley is not at a monopoly price, and consequently it can be raised in proportion to the taxes that may be imposed upon it. This opinion of Mr. Buchanan of the effects of a tax on malt appears to me to be in direct contradiction to the opinion he has given of a similar tax, a tax on bread. " A tax on bread will be ultimately paid, not by a rise of price, but by a reduction of rent." [1] If a tax on malt would raise the price of beer, a tax on bread must raise the price of bread.

The following argument of M. Say is founded on the same views as Mr. Buchanan's: " The quantity of wine or corn which a piece of land will produce will remain nearly the same, what-ever may be the tax with which it is charged. The tax may take away a half, or even three-fourths of its net produce, or of its rent, if you please, yet the land would nevertheless be culti-vated for the half or the quarter not absorbed by the tax. The rent, that is to say, the landlord's share, would merely be somewhat lower. The reason of this will be perceived if we consider that, in the case supposed, the quantity of produce

[1] Vol. iii. p. 355.

obtained from the land and sent to market will remain nevertheless the same. On the other hand, the motives on which
the demand for the produce is founded continue also the
same.

" Now, if the quantity of produce supplied, and the quantity
demanded, necessarily continue the same, notwithstanding the
establishment or the increase of the tax, the price of that produce
will not vary; and if the price do not vary, the consumer will
not pay the smallest portion of this tax.

" Will it be said that the farmer, he who furnishes labour and
capital, will, jointly with the landlord, bear the burden of this
tax?—certainly not; because the circumstance of the tax has not
diminished the number of farms to be let, nor increased the
number of farmers. Since, in this instance also, the supply and
demand remain the same, the rent of farms must also remain
the same. The example of the manufacturer of salt, who can
only make the consumers pay a portion of the tax, and that
of the landlord, who cannot reimburse himself in the smallest
degree, prove the error of those who maintain, in opposition to
the economists, that all taxes fall ultimately on the consumer."
—Vol. ii. p. 338.

If the tax " took away half, or even three-fourths of the net
produce of the land," and the price of produce did not rise, how
could those farmers obtain the usual profits of stock who paid
very moderate rents, having that quality of land which required
a much larger proportion of labour to obtain a given result than
land of a more fertile quality? If the whole rent were remitted,
they would still obtain lower profits than those in other trades,
and would therefore not continue to cultivate their land, unless
they could raise the price of its produce. If the tax fell on the
farmers, there would be fewer farmers disposed to hire farms;
if it fell on the landlord, many farms would not be let at all, for
they would afford no rent. But from what fund would those
pay the tax who produce corn without paying any rent? It is
quite clear that the tax must fall on the consumer. How would
such land as M. Say describes in the following passage pay a tax
of one-half or three-fourths of its produce?

" We see in Scotland poor lands thus cultivated by the proprietor, and which could be cultivated by no other person.
Thus, too, we see in the interior provinces of the United States
vast and fertile lands, the revenue of which, alone, would not be
sufficient for the maintenance of the proprietor. These lands

are cultivated nevertheless, but it must be by the proprietor himself, or, in other words, he must add to the rent, which is little or nothing, the profits of his capital and industry, to enable him to live in competence. It is well known that land, though cultivated, yields no revenue to the landlord when no farmer will be willing to pay a rent for it: which is a proof that such land will give only the profits of the capital, and of the industry necessary for its cultivation."—Say, vol. ii. p. 127.

CHAPTER XVIII

POOR RATES

WE have seen that taxes on raw produce, and on the profits of the farmer, will fall on the consumer of raw produce; since, unless he had the power of remunerating himself by an increase of price, the tax would reduce his profits below the general level of profits, and would urge him to remove his capital to some other trade. We have seen, too, that he could not, by deducting it from his rent, transfer the tax to his landlord; because that farmer who paid no rent would, equally with the cultivator of better land, be subject to the tax, whether it were laid on raw produce or on the profits of the farmer. I have also attempted to show that if a tax were general, and affected equally all profits, whether manufacturing or agricultural, it would not operate either on the price of goods or raw produce, but would be immediately, as well as ultimately, paid by the producers. A tax on rent, it has been observed, would fall on the landlord only, and could not by any means be made to devolve on the tenant.

The poor rate is a tax which partakes of the nature of all these taxes, and, under different circumstances, falls on the consumer of raw produce and goods, on the profits of stock, and on the rent of land. It is a tax which falls with peculiar weight on the profits of the farmer, and therefore may be considered as affecting the price of raw produce. According to the degree in which it bears on manufacturing and agricultural profits equally, it will be a general tax on the profits of stock, and will occasion no alteration in the price of raw produce and manufactures. In proportion to the farmer's inability to remunerate himself, by raising the price of raw produce for that portion of the tax which peculiarly affects him, it will be a tax on rent and will be paid by the landlord. To know, then, the operation of the poor rate at any particular time, we must ascertain whether at that time it affects in an equal or an unequal degree the profits of the farmer and manufacturer; and also whether the circumstances be such as to afford to the farmer the power of raising the price of raw produce.

The poor rates are professed to be levied on the farmer in

proportion to his rent; and, accordingly, the farmer who paid a very small rent, or no rent at all, should pay little or no tax. If this were true, poor rates, as far as they are paid by the agricultural class, would entirely fall on the landlord, and could not be shifted to the consumer of raw produce. But I believe that it is not true; the poor rate is not levied according to the rent which a farmer actually pays to his landlord; it is proportioned to the annual value of his land, whether that annual value be given to it by the capital of the landlord or of the tenant.

If two farmers rented land of two different qualities in the same parish, the one paying a rent of £100 per annum for 50 acres of the most fertile land, and the other the same sum of £100 for 1000 acres of the least fertile land, they would pay the same amount of poor rates, if neither of them attempted to improve the land; but if the farmer of the poor land, presuming on a very long lease, should be induced, at a great expense, to improve the productive powers of his land, by manuring, draining, fencing, etc., he would contribute to the poor rates, not in proportion to the actual rent paid to the landlord, but to the actual annual value of the land. The rate might equal or exceed the rent; but whether it did or not, no part of this rate would be paid by the landlord. It would have been previously calculated upon by the tenant; and if the price of produce were not sufficient to compensate him for all his expenses, together with this additional charge for poor rates, his improvements would not have been undertaken. It is evident, then, that the tax in this case is paid by the consumer; for if there had been no rate, the same improvements would have been undertaken, and the usual and general rate of profits would have been obtained on the stock employed with a lower price of corn.

Nor would it make the slightest difference in this question if the landlord had made these improvements himself, and had in consequence raised his rent from £100 to £500; the rate would be equally charged to the consumer; for whether the landlord should expend a large sum of money on his land would depend on the rent, or what is called rent, which he would receive as a remuneration for it; and this again would depend on the price of corn, or other raw produce, being sufficiently high, not only to cover this additional rent, but also the rate to which the land would be subject. If at the same time all manufacturing capital contributed to the poor rates in the same proportion as the capital expended by the farmer or landlord in improving the land, then it would no longer be a partial tax on the profits of the farmer's or landlord's capital, but a tax on the capital of all producers; and, therefore, it could no longer be shifted either on the consumer of raw produce or on the landlord. The farmer's profits would feel the effect of the rate no more than those of the manufacturer; and the former could not, any more than the latter, plead it as a reason for an advance in the price of his commodity. It is not the absolute but the relative fall of profits which prevents capital from being employed in any particular trade: it is the difference of profit which sends capital from one employment to another.

It must be acknowledged, however, that in the actual state of the poor rates, a much larger amount falls on the farmer than on the manufacturer, in proportion to their respective profits; the farmer being rated according to the actual productions which he obtains, the manufacturer only according to the value of the buildings in which he works, without any regard to the value of the machinery, labour, or stock which he may employ. From this circumstance it follows that the farmer will be enabled to raise the price of his produce by this whole difference. For since the tax falls unequally, and peculiarly on his profits, he would have less motive to devote his capital to the land than to employ it in some other trade, were not the price of raw produce raised. If, on the contrary, the rate had fallen with greater weight on the manufacturer than on the farmer, he would have been enabled to raise the price of his goods by the amount of the difference, for the same reason that the farmer under similar circumstances could raise the price of raw produce. In a society, therefore, which is extending its agriculture, when poor rates fall with peculiar weight on the land, they will be paid partly by the employers of capital in a diminution of the profits of stock, and partly by the consumer of raw produce in its increased price. In such a state of things, the tax may, under some circumstances, be even advantageous rather than injurious to landlords; for if the tax paid by the cultivator of the worst land be higher in proportion to the quantity of produce obtained than that paid by the farmers of the more fertile lands, the rise in the price of corn, which will extend to all corn, will more than compensate the latter for the tax. This advantage will remain with them during the continuance of their leases, but it will afterwards be transferred to their landlords. This, then, would be the effect of poor rates in an advancing society; but in a stationary, or in a retrograde country, so far as capital could not be withdrawn from the land, if a further rate were levied for the support of the poor, that part of it which fell on agriculture would be paid, during the current leases, by the farmers; but, at the expiration of those leases it would almost wholly fall on the landlords. The farmer, who, during his former lease, had expended his capital in improving his land, if it were still in his own lands, would be rated for this new tax according to the new value which the land had acquired by its improvement, and this amount he would be obliged to pay during his lease, although his profits might thereby be reduced below the general rate of profits; for the capital which he has expended may be so incorporated with the land that it cannot be removed from it. If, indeed, he or his landlord (should it have been expended by him) were able to remove this capital, and thereby reduce the annual value of the land, the rate would proportionably fall; and as the produce would at the same time be diminished, its price would rise; he would be compensated for the tax by charging it to the consumer, and no part would fall on rent; but this is impossible, at least with respect to some proportion of the capital, and consequently in that proportion the tax will be paid by the farmers during their leases, and by landlords at their expiration. This additional tax, if it fell with peculiar severity on manufacturers, which it does not, would, under such circumstances, be added to the price of their goods; for there can be no reason why their profits should be reduced below the general rate of profits when their capitals might be easily removed to agriculture.[1]

[1] In a former part of this work I have noticed the difference between rent, properly so called, and the remuneration paid to the landlord under that name for the advantages which the expenditure of his capital has procured to his tenant; but I did not perhaps sufficiently distinguish the difference which would arise from the different modes in which this capital might be applied. As a part of this capital, when once expended in the improvement of a farm, is inseparably amalgamated with the land, and tends to increase its productive powers, the remuneration paid to the landlord for its use is strictly of the nature of rent, and is subject to all the laws of rent. Whether the improvement be made at the expense of the landlord or the tenant, it will not be undertaken in the first instance unless there is a strong probability that the return will at least be equal to the profit that can be made by the disposition of any other equal capital; but when once made, the return obtained will ever after be wholly of the nature of rent, and will be subject to all the variations of rent. Some of these expenses, however, only give advantages to the land for a limited period, and do not add permanently to its productive powers: being bestowed on buildings, and other perishable improvements, they require to be constantly renewed, and therefore do not obtain for the landlord any permanent addition to his real rent.

CHAPTER XIX

ON SUDDEN CHANGES IN THE CHANNELS OF TRADE

A GREAT manufacturing country is peculiarly exposed to temporary reverses and contingencies, produced by the removal of capital from one employment to another. The demands for the produce of agriculture are uniform; they are not under the influence of fashion, prejudice, or caprice. To sustain life, food is necessary, and the demand for food must continue in all ages and in all countries. It is different with manufactures; the demand for any particular manufactured commodity is subject, not only to the wants, but to the tastes and caprice of the purchasers. A new tax, too, may destroy the comparative advantage which a country before possessed in the manufacture of a particular commodity; or the effects of war may so raise the freight and insurance on its conveyance, that it can no longer enter into competition with the home manufacture of the country to which it was before exported. In all such cases, considerable distress, and no doubt some loss, will be experienced by those who are engaged in the manufacture of such commodities; and it will be felt, not only at the time of the change, but through the whole interval during which they are removing their capitals, and the labour which they can command, from one employment to another.

Nor will distress be experienced in that country alone where such difficulties originate, but in the countries to which its commodities were before exported. No country can long import, unless it also exports, or can long export unless it also imports. If, then, any circumstance should occur which should permanently prevent a country from importing the usual amount of foreign commodities, it will necessarily diminish the manufacture of some of those commodities which were usually exported; and although the total value of the productions of the country will probably be but little altered, since the same capital will be employed, yet they will not be equally abundant and cheap; and considerable distress will be experienced through the change of employments. If, by the employment of £10,000 in the manufacture of cotton goods for exportation, we imported annually 3000 pair of silk stockings of the value of £2000, and by the interruption of foreign trade we should be obliged to withdraw this capital from the manufacture of cotton, and employ it ourselves in the manufacture of stockings, we should still obtain stockings of the value of £2000, provided no part of the capital were destroyed; but instead of having 3000 pair, we might only have 2500. In the removal of the capital from the cotton to the stocking trade, much distress might be experienced, but it would not considerably impair the value of the national property, although it might lessen the quantity of our annual productions.[1]

The commencement of war after a long peace, or of peace after a long war, generally produces considerable distress in trade. It changes in a great degree the nature of the employments to which the respective capitals of countries were before devoted; and during the interval while they are settling in the situations which new circumstances have made the most beneficial, much fixed capital is unemployed, perhaps wholly lost, and labourers are without full employment. The duration of this distress will be longer or shorter according to the strength of that disinclination which most men feel to abandon that employment of their capital to which they have long been accustomed. It is often protracted, too, by the restrictions and prohibitions to which the absurd jealousies which prevail between the different states of the commercial commonwealth give rise.

The distress which proceeds from a revulsion of trade is often mistaken for that which accompanies a diminution of the national capital and a retrograde state of society; and it would perhaps be difficult to point out any marks by which they may be accurately distinguished.

When, however, such distress immediately accompanies a

[1] " Commerce enables us to obtain a commodity in the place where it is to be found, and to convey it to another where it is to be consumed; it therefore gives us the power of increasing the value of the commodity, by the whole difference between its price in the first of these places and its price in the second."—M. Say, p. 458, vol. ii.—True, but how is this additional value given to it? By adding to the cost of production, first, the expenses of conveyance; secondly, the profit on the advances of capital made by the merchant. The commodity is only more valuable for the same reasons that every other commodity may become more valuable, because more labour is expended on its production and conveyance before it is purchased by the consumer. This must not be mentioned as one of the advantages of commerce. When the subject is more closely examined, it will be found that the whole benefits of commerce resolve themselves into the means which it gives us of acquiring, not more valuable objects, but more useful ones.

change from war to peace, our knowledge of the existence of such a cause will make it reasonable to believe that the funds for the maintenance of labour have rather been diverted from their usual channel than materially impaired, and that, after temporary suffering, the nation will again advance in prosperity. It must be remembered, too, that the retrograde condition is always an unnatural state of society. Man from youth grows to manhood, then decays, and dies; but this is not the progress of nations. When arrived to a state of the greatest vigour, their further advance may indeed be arrested, but their natural tendency is to continue for ages to sustain undiminished their wealth and their population.

In rich and powerful countries, where large capitals are invested in machinery, more distress will be experienced from a revulsion in trade than in poorer countries where there is proportionally a much smaller amount of fixed, and a much larger amount of circulating capital, and where consequently more work is done by the labour of men. It is not so difficult to withdraw a circulating as a fixed capital from any employment in which it may be engaged. It is often impossible to divert the machinery which may have been erected for one manufacture to the purposes of another; but the clothing, the food, and the lodging of the labourer in one employment may be devoted to the support of the labourer in another; or the same labourer may receive the same food, clothing, and lodging, whilst his employment is changed. This, however, is an evil to which a rich nation must submit; and it would not be more reasonable to complain of it than it would be in a rich merchant to lament that his ship was exposed to the dangers of the sea, whilst his poor neighbour's cottage was safe from all such hazard.

From contingencies of this kind, though in an inferior degree, even agriculture is not exempted. War, which, in a commercial country, interrupts the commerce of states, frequently prevents the exportation of corn from countries where it can be produced with little cost to others not so favourably situated. Under such circumstances an unusual quantity of capital is drawn to agriculture, and the country which before imported becomes independent of foreign aid. At the termination of the war, the obstacles to importation are removed, and a competition destructive to the home-grower commences, from which he is unable to withdraw without the sacrifice of a great part of his capital. The best policy of the state would be to lay a tax, decreasing in amount from time to time, on the importation of foreign corn, for a limited number of years, in order to afford to the home-grower an opportunity to withdraw his capital gradually from the land.[1] In so doing, the country might not be making the most advantageous distribution of its capital, but the temporary tax to which it was subjected would be for the advantage of a particular class, the distribution of whose capital was highly useful in procuring a supply of food when importation was stopped. If such exertions in a period of emergency were followed by a risk of ruin on the termination of the difficulty, capital would shun such an employment. Besides the usual profits of stock, farmers would expect to be compensated for the risk which they incurred of a sudden influx of corn; and, therefore, the price to the consumer, at the seasons when he most required a supply, would be enhanced, not only by the superior cost of growing corn at home, but also by the insurance which he would have to pay in the price for the peculiar risk to which this employment of capital was exposed. Notwithstanding, then, that it would be more productive of wealth to the country, at whatever sacrifice of capital it might be done, to allow the importation of cheap corn, it would, perhaps, be advisable to charge it with a duty for a few years.

In examining the question of rent, we found that, with every increase in the supply of corn, and with the consequent fall of its price, capital would be withdrawn from the poorer land, and land of a better description, which would then pay no rent,

[1] In the last volume of the supplement to the *Encyclopædia Britannica*, article " Corn Laws and Trade," are the following excellent suggestions and observations:—" If we shall at any future period think of retracing our steps, in order to give time to withdraw capital from the cultivation of our poor soils, and to invest it in more lucrative employments, a gradually diminishing scale of duties may be adopted. The price at which foreign grain should be admitted duty free may be made to decrease from 80s., its present limit, by 4s. or 5s. per quarter annually till it reaches 50s., when the ports could safely be thrown open, and the restrictive system be for ever abolished. When this happy event shall have taken place, it will be no longer necessary to force nature. The capital and enterprise of the country will be turned into those departments of industry in which our physical situation, national character, or political institutions fit us to excel. The corn of Poland and the raw cotton of Carolina will be exchanged for the wares of Birmingham and the muslins of Glasgow. The genuine commercial spirit, that which permanently secures the prosperity of nations, is altogether inconsistent with the dark and shallow policy of monopoly. The nations of the earth are like provinces of the same kingdom—a free and unfettered intercourse is alike productive of general and of local advantage." The whole article is well worthy of attention; it is very instructive, is ably written, and shows that the author is completely master of the subject.

would become the standard by which the natural price of corn would be regulated. At £4 per quarter, land of an inferior quality, which may be designated by No. 6, might be cultivated; at £3 10s., No. 5; at £3, No. 4, and so on. If corn, in consequence of permanent abundance, fell to £3 10s., the capital employed on No. 6 would cease to be employed; for it was only when corn was at £4 that it could obtain the general profits, even without paying rent: it would, therefore, be withdrawn to manufacture those commodities with which all the corn grown on No. 6 would be purchased and imported. In this employment it would necessarily be more productive to its owner, or it would not be withdrawn from the other; for if he could not obtain more corn by purchasing it with a commodity which he manufactured than he got from the land for which he paid no rent, its price could not be under £4.

It has, however, been said, that capital cannot be withdrawn from the land; that it takes the form of expenses which cannot be recovered, such as manuring, fencing, draining, etc., which are necessarily inseparable from the land. This is in some degree true; but that capital which consists of cattle, sheep, hay and corn ricks, carts, etc., may be withdrawn; and it always becomes a matter of calculation whether these shall continue to be employed on the land, notwithstanding the low price of corn, or whether they shall be sold, and their value transferred to another employment.

Suppose, however, the fact to be as stated, and that no part of the capital could be withdrawn;[1] the farmer would continue to raise corn, and precisely the same quantity, too, at whatever price it might sell; for it could not be his interest to produce

[1] Whatever capital becomes fixed on the land must necessarily be the landlord's, and not the tenant's, at the expiration of the lease. Whatever compensation the landlord may receive for this capital on re-letting his land will appear in the form of rent; but no rent will be paid if, with a given capital, more corn can be obtained from abroad than can be grown on this land at home. If the circumstances of the society should require corn to be imported, and 1000 quarters can be obtained by the employment of a given capital, and if this land, with the employment of the same capital, will yield 1100 quarters, 100 quarters will necessarily go to rent; but if 1200 can be got from abroad, then this land will go out of cultivation, for it will not then yield even the general rate of profit. But this is no disadvantage, however great the capital may have been that had been expended on the land. Such capital is spent with a view to augment the produce—that, it should be remembered, is the end; of what importance, then, can it be to the society whether half its capital be sunk in value, or even annihilated, if they obtain a great annual quantity of production? Those who deplore the loss of capital in this case are for sacrificing the end to the means.

less, and if he did not so employ his capital, he would obtain from it no return whatever. Corn would not be imported, because he would sell it lower than £3 10s. rather than not sell it at all, and by the supposition the importer could not sell it under that price. Although, then, the farmers, who cultivated land of this quality, would undoubtedly be injured by the fall in the exchangeable value of the commodity which they produced—how would the country be affected? We should have precisely the same quantity of every commodity produced, but raw produce and corn would sell at a much cheaper price. The capital of a country consists of its commodities, and as these would be the same as before, reproduction would go on at the same rate. This low price of corn would, however, only afford the usual profits of stock to the land No. 5, which would then pay no rent, and the rent of all better land would fall: wages would also fall, and profits would rise.

However low the price of corn might fall, if capital could not be removed from the land, and the demand did not increase, no importation would take place, for the same quantity as before would be produced at home. Although there would be a different division of the produce, and some classes would be benefited and others injured, the aggregate of production would be precisely the same, and the nation collectively would neither be richer nor poorer.

But there is this advantage always resulting from a relatively low price of corn—that the division of the actual production is more likely to increase the fund for the maintenance of labour, inasmuch as more will be allotted, under the name of profit, to the productive class — a less, under the name rent, to the unproductive class.

This is true, even if the capital cannot be withdrawn from the land, and must be employed there, or not be employed at all; but if great part of the capital can be withdrawn, as it evidently could, it will be only withdrawn when it will yield more to the owner by being withdrawn than by being suffered to remain where it was; it will only be withdrawn then, when it can elsewhere be employed more productively both for the owner and the public. He consents to sink that part of his capital which cannot be separated from the land, because with that part which he can take away he can obtain a greater value, and a greater quantity of raw produce, than by not sinking this part of the capital. His case is precisely similar to that of a man who has erected machinery in his manufactory at a great expense, machinery which is afterwards so much improved upon by more modern inventions that the commodities manufactured by him very much sink in value. It would be entirely a matter of calculation with him whether he should abandon the old machinery, and erect the more perfect, *losing all the value of the old*, or continue to avail himself of its comparatively feeble powers. Who, under such circumstances, would exhort him to forego the use of the better machinery, because it would deteriorate or annihilate the value of the old? Yet, this is the argument of those who would wish us to prohibit the importation of corn, because it will deteriorate or annihilate that part of the capital of the farmer which is for ever sunk in land. They do not see that the end of all commerce is to increase production, and that, by increasing production, though you may occasion partial loss, you increase the general happiness. To be consistent, they should endeavour to arrest all improvements in agriculture and manufactures, and all inventions of machinery; for, though these contribute to general abundance, and therefore to the general happiness, they never fail, at the moment of their introduction, to deteriorate or annihilate the value of a part of the existing capital of farmers and manufacturers.[1]

Agriculture, like all the other trades, and particularly in a commercial country, is subject to a reaction, which, in an opposite direction, succeeds the action of a strong stimulus. Thus, when war interrupts the importation of corn, its consequent high price attracts capital to the land, from the large profits which such an employment of it affords; this will probably cause more capital to be employed, and more raw produce to be brought to market than the demands of the country require. In such case, the price of corn will fall from the effects of a glut, and much agricultural distress will be produced, till the average supply is brought to a level with the average demand.

[1] Among the most able of the publications on the impolicy of restricting the importation of corn may be classed Major Torrens' *Essay on the External Corn Trade.* His arguments appear to me to be unanswered, and to be unanswerable.

CHAPTER XX

" A MAN is rich or poor," says Adam Smith, " according to the degree in which he can afford to enjoy the necessaries, conveniences, and amusements of human life."

Value, then, essentially differs from riches, for value depends not on abundance, but on the difficulty or facility of production. The labour of a million of men in manufactures will always produce the same value, but will not always produce the same riches. By the invention of machinery, by improvements in skill, by a better division of labour, or by the discovery of new markets, where more advantageous exchanges may be made, a million of men may produce double or treble the amount of riches, of " necessaries, conveniences, and amusements," in one state of society that they could produce in another, but they will not on that account add anything to value; for everything rises or falls in value in proportion to the facility or difficulty of producing it, or, in other words, in proportion to the quantity of labour employed on its production. Suppose, with a given capital, the labour of a certain number of men produced 1000 pair of stockings, and that by inventions in machinery the same number of men can produce 2000 pair, or that they can continue to produce 1000 pair, and can produce besides 500 hats; then the value of the 2000 pair of stockings, or of the 1000 pair of stockings and 500 hats, will be neither more nor less than that of the 1000 pair of stockings before the introduction of machinery; for they will be the produce of the same quantity of labour. But the value of the general mass of commodities will nevertheless be diminished; for, although the value of the increased quantity produced in consequence of the improvement will be the same exactly as the value would have been of the less quantity that would have been produced, had no improvement taken place, an effect is also produced on the portion of goods still unconsumed, which were manufactured previously to the improvement; the value of those goods will be reduced, inasmuch as they must fall to the level, quantity for quantity, of the goods produced under all the advantages of the improvement: and the society will, notwithstanding the increased quantity of commodities, notwithstanding its augmented riches, and its augmented means of enjoyment, have a less amount of value. By constantly increasing the facility of production, we constantly diminish the value of some of the commodities before produced, though by the same means we not only add to the national riches, but also to the power of future production. Many of the errors in political economy have arisen from errors on this subject, from considering an increase of riches, and an increase of value, as meaning the same thing, and from unfounded notions as to what constituted a standard measure of value. One man considers money as a standard of value, and a nation grows richer or poorer, according to him, in proportion as its commodities of all kinds can exchange for more or less money. Others represent money as a very convenient medium for the purpose of barter, but not as a proper measure by which to estimate the value of other things; the real measure of value according to them is corn,[1] and a country is rich or poor according as its commodities will exchange for more or less corn.[2] There are others again who consider a country rich or poor according to the quantity of labour that it can purchase. But why should gold, or corn, or labour, be the standard measure of value, more than coals or iron?—more than cloth, soap, candles, and the other necessities of the labourer?—why, in short, should any commodity, or all commodities together, be the standard, when such a standard is itself subject to fluctuations in value? Corn, as well as gold, may from difficulty or facility of production vary 10, 20, or 30 per cent. relatively to other things; why should we always say that it is those other things which have varied, and not the corn? That commodity is alone invariable which at all times requires the same sacrifice of toil and labour to produce it. Of such a commodity we have no knowledge, but we may hypothetically argue and speak about it as if we had; and may improve our knowledge of the science

[1] Adam Smith says, " that the difference between the real and the nominal price of commodities and labour is not a matter of mere speculation, but may sometimes be of considerable use in practice." I agree with him; but the real price of labour and commodities is no more to be ascertained by their price in goods, Adam Smith's real measure, than by their price in gold and silver, his nominal measure. The labourer is only paid a really high price for his labour when his wages will purchase the produce of a great deal of labour.

[2] In vol. i. p. 108, M. Say infers that silver is now of the same value as in the reign of Louis XIV., " because the same quantity of silver will buy the same quantity of corn."

by showing distinctly the absolute inapplicability of all the standards which have been hitherto adopted. But supposing either of these to be a correct standard of value, still it would not be a standard of riches, for riches do not depend on value. A man is rich or poor according to the abundance of necessaries and luxuries which he can command; and whether the exchangeable value of these for money, for corn, or for labour be high or low, they will equally contribute to the enjoyment of their possessor. It is through confounding the ideas of value and wealth, or riches, that it has been asserted that by diminishing the quantity of commodities, that is to say, of the necessaries, conveniences, and enjoyments of human life, riches may be increased. If value were the measure of riches, this could not be denied, because by scarcity the value of commodities is raised; but if Adam Smith be correct, if riches consist in necessaries and enjoyments, then they cannot be increased by a diminution of quantity.

It is true that the man in possession of a scarce commodity is richer, if by means of it he can command more of the necessaries and enjoyments of human life; but as the general stock out of which each man's riches are drawn is diminished in quantity by all that any individual takes from it, other men's shares must necessarily be reduced in proportion as this favoured individual is able to appropriate a greater quantity to himself.

Let water become scarce, says Lord Lauderdale, and be exclusively possessed by an individual, and you will increase his riches, because water will then have value; and if wealth be the aggregate of individual riches, you will by the same means also increase wealth. You undoubtedly will increase the riches of this individual, but inasmuch as the farmer must sell a part of his corn, the shoemaker a part of his shoes, and all men give up a portion of their possessions for the sole purpose of supplying themselves with water, which they before had for nothing, they are poorer by the whole quantity of commodities which they are obliged to devote to this purpose, and the proprietor of water is benefited precisely by the amount of their loss. The same quantity of water, and the same quantity of commodities, are enjoyed by the whole society, but they are differently distributed. This is, however, supposing rather a monopoly of water than a scarcity of it. If it should be scarce, then the riches of the country and of individuals would be actually diminished, inasmuch as it would be deprived of a portion of one of its enjoyments. The farmer would not only have less corn to exchange for the other commodities which might be necessary or desirable to him, but he, and every other individual, would be abridged in the enjoyment of one of the most essential of their comforts. Not only would there be a different distribution of riches, but an actual loss of wealth.

It may be said, then, of two countries possessing precisely the same quantity of all the necessaries and comforts of life, that they are equally rich, but the value of their respective riches would depend on the comparative facility or difficulty with which they were produced. For if an improved piece of machinery should enable us to make two pair of stockings instead of one, without additional labour, double the quantity would be given in exchange for a yard of cloth. If a similar improvement be made in the manufacture of cloth, stockings and cloth will exchange in the same proportions as before, but they will both have fallen in value; for in exchanging them for hats, for gold, or other commodities in general, twice the former quantity must be given. Extend the improvement to the production of gold, and every other commodity, and they will all regain their former proportions. There will be double the quantity of commodities annually produced in the country, and therefore the wealth of the country will be doubled, but this wealth will not have increased in value.

Although Adam Smith has given the correct description of riches which I have more than once noticed, he afterwards explains them differently, and says, " that a man must be rich or poor according to the quantity of labour which he can afford to purchase." Now, this description differs essentially from the other, and is certainly incorrect; for suppose the mines were to become more productive, so that gold and silver fell in value, from the greater facility of their production; or that velvets were to be manufactured with so much less labour than before, that they fell to half their former value; the riches of all those who purchased those commodities would be increased; one man might increase the quantity of his plate, another might buy double the quantity of velvet; but with the possession of this additional plate and velvet, they could employ no more labour than before; because, as the exchangeable value of velvet and of plate would be lowered, they must part with proportionally more of these species of riches to purchase a day's labour. Riches, then, cannot be estimated by the quantity of labour which they can purchase.

From what has been said, it will be seen that the wealth of a country may be increased in two ways: it may be increased by employing a greater portion of revenue in the maintenance of productive labour, which will not only add to the quantity, but to the value of the mass of commodities; or it may be increased, without employing any additional quantity of labour, by making the same quantity more productive, which will add to the abundance, but not to the value of commodities.

In the first case, a country would not only become rich, but the value of its riches would increase. It would become rich by parsimony—by diminishing its expenditure on objects of luxury and enjoyment, and employing those savings in reproduction.

In the second case, there will not necessarily be either any diminished expenditure on luxuries and enjoyments, or any increased quantity of productive labour employed, but, with the same labour, more would be produced; wealth would increase, but not value. Of these two modes of increasing wealth, the last must be preferred, since it produces the same effect without the privation and diminution of enjoyments which can never fail to accompany the first mode. Capital is that part of the wealth of a country which is employed with a view to future production, and may be increased in the same manner as wealth. An additional capital will be equally efficacious in the production of future wealth, whether it be obtained from improvements in skill and machinery, or from using more revenue reproductively; for wealth always depends on the quantity of commodities produced, without any regard to the facility with which the instruments employed in production may have been procured. A certain quantity of clothes and provisions will maintain and employ the same number of men, and will therefore procure the same quantity of work to be done, whether they be produced by the labour of 100 or 200 men; but they will be of twice the value if 200 have been employed on their production.

M. Say, notwithstanding the corrections he has made in the fourth and last edition of his work, *Traité d'Economie Politique*, appears to me to have been singularly unfortunate in his definition of riches and value. He considers these two terms as synonymous, and that a man is rich in proportion as he increases the value of his possessions, and is enabled to command an abundance of commodities. "The value of incomes is then increased," he observes, "if they can procure, it does not signify by what means, a greater quantity of products." According to M. Say, if the difficulty of producing cloth were to double, and consequently cloth was to exchange for double the quantity of the commodities for which it is exchanged before, it would be doubled in value, to which I give my fullest assent; but if there were any peculiar facility in producing the commodities, and no increased difficulty in producing cloth, and cloth should in consequence exchange as before for double the quantity of commodities, M. Say would still say that cloth had doubled in value, whereas, according to my view of the subject, he should say, that cloth retained its former value, and those particular commodities had fallen to half their former value. Must not M. Say be inconsistent with himself when he says that, by facility of production, two sacks of corn may be produced by the same means that one was produced before, and that each sack will therefore fall to half its former value, and yet maintain that the clothier who exchanges his cloth for two sacks of corn will obtain double the value he before obtained, when he could only get one sack in exchange for his cloth. If two sacks be of the value that one was of before, he evidently obtains the same value and no more—he gets, indeed, double the quantity of riches—double the quantity of utility—double the quantity of what Adam Smith calls value in use, but not double the quantity of value, and therefore M. Say cannot be right in considering value, riches, and utility to be synonymous. Indeed, there are many parts of M. Say's work to which I can confidently refer in support of the doctrine which I maintain respecting the essential difference between value and riches, although it must be confessed that there are also various other passages in which a contrary doctrine is maintained. These passages I cannot reconcile, and I point them out by putting them in opposition to each other, that M. Say may, if he should do me the honour to notice these observations in any future edition of his work, give such explanations of his views as may remove the difficulty which many others, as well as myself, feel in our endeavours to expound them.

1. In the exchange of two products, we only in fact exchange the productive services which have served to create them.
p. 504

2. There is no real dearness but that which arises from the cost of production. A thing really dear is that which costs much in producing . . p. 497

3. The value of all the productive services that must be consumed to create a product constitute the cost of production of that product . p. 505

4. It is utility which determines the demand for a commodity, but it is the cost of its production which limits the extent of its demand. When its utility does not elevate its value to the level of the cost of production, the thing is not worth what it cost; it is a

proof that the productive services might be employed to
create a commodity of a
superior value. The possessors of productive funds, that
is to say, those who have the
disposal of labour, of capital or
land, are perpetually occupied
in comparing the cost of production with the value of the
things produced, or, which
comes to the same thing, in
comparing the value of different commodities with each
other; because the cost of production is nothing else but the
value of productive services,
consumed in forming a production; and the value of a
productive service is nothing
else than the value of the commodity, which is the result.
The value of a commodity, the
value of a productive service,
the value of the cost of production, are all, then, similar
values, when everything is left
to its natural course.

5. The value of incomes is then
increased, if they can procure
(it does not signify by what
means) a greater quantity of
products.

6. Price is the measure of the
value of things, and their value
is the measure of their utility.
Vol. 2, p. 4

7. Exchanges made freely show
at the time, in the place, and

in the state of society in which
we are the value which men
attach to the things ex-
changed   .   .   p. 466

8. To produce, is to create value,
by giving or increasing the
utility of a thing, and thereby
establishing a demand for it,
which is the first cause of its
value   .   .   Vol. 2, p. 487

9. Utility being created, constitutes a product. The exchangeable value which results
is only the measure of this
utility, the measure of the production which has taken place.
p. 490

10. The utility which people of a
particular country find in a
product can no otherwise be
appreciated than by the price
which they give for it . p. 502

11. This price is the measure of the
utility which it has in the
judgment of men; of the satisfaction which they derive from
consuming it, because they
would not prefer consuming
this utility, if for the price
which it cost they could acquire
a utility which would give
them more satisfaction . p. 506

12. The quantity of all other commodities which a person can
immediately obtain in exchange
for the commodity of which he
wishes to dispose, is at all times
a value not to be disputed.
Vol. 2, p. 4

If there is no real dearness but that which arises from cost of
production (see 2) how can a commodity be said to rise in value
(see 5), if its cost of production be not increased? and merely
because it will exchange for more of a cheap commodity—for
more of a commodity the cost of production of which has
diminished? When I give 2000 times more cloth for a pound
of gold than I give for a pound of iron, does it prove that I
attach 2000 times more utility to gold than I do to iron? certainly not; it proves only as admitted by M. Say (see 4), that the
cost of production of gold is 2000 times greater than the cost of
production of iron. If the cost of production of the two metals
were the same, I should give the same price for them; but if
utility were the measure of value, it is probable I should give
more for the iron. It is the competition of the producers " who

are perpetually employed in comparing the cost of production with the value of the thing produced " (*see* 4) which regulates the value of different commodities. If, then, I give one shilling for a loaf, and 21 shillings for a guinea, it is no proof that this in my estimation is the comparative measure of their utility.

In No. 4, M. Say maintains, with scarcely any variation, the doctrine which I hold concerning value. In his productive services he includes the services rendered by land, capital, and labour; in mine I include only capital and labour, and wholly exclude land. Our difference proceeds from the different view which we take of rent: I always consider it as the result of a partial monopoly, never really regulating price, but rather as the effect of it. If all rent were relinquished by landlords, I am of opinion that the commodities produced on the land would be no cheaper, because there is always a portion of the same commodities produced on land for which no rent is or can be paid, as the surplus produce is only sufficient to pay the profits of stock.

To conclude, although no one is more disposed than I am to estimate highly the advantage which results to all classes of consumers from the real abundance and cheapness of commodities, I cannot agree with M. Say in estimating the value of a commodity by the abundance of other commodities for which it will exchange; I am of the opinion of a very distinguished writer, M. Destutt de Tracy, who says that, " To measure any one thing is to compare it with a determinate quantity of that same thing which we take for a standard of comparison, for unity. To measure, then, to ascertain a length, a weight, a value, is to find how many times they contain metres, grammes, francs, in a word, unities of the same description." A franc is not a measure of value for any thing, but for a quantity of the same metal of which francs are made, unless francs, and the thing to be measured, can be referred to some other measure which is common to both. This, I think, they can be, for they are both the result of labour; and, therefore, labour is a common measure, by which their real as well as their relative value may be estimated. This also, I am happy to say, appears to be M. Destutt de Tracy's opinion.[1] He says, " As it is certain that our physical and moral faculties are alone our original riches, the employment

[1] *Elemens d'Ideologie*, vol. iv. p. 99.—In this work M. de Tracy has given a useful and an able treatise on the general principles of Political Economy, and I am sorry to be obliged to add that he supports, by his authority, the definitions which M. Say has given of the words " value," " riches," and " utility."

of those faculties, labour of some kind, is our only original
treasure, and that it is always from this employment that all
those things are created which we call riches, those which are
the most necessary as well as those which are the most purely
agreeable. It is certain too, that all those things only represent
the labour which has created them, and if they have a value, or
even two distinct values, they can only derive them from that
of the labour from which they emanate."

M. Say, in speaking of the excellences and imperfections of the
great work of Adam Smith, imputes to him, as an error, that
" he attributes to the labour of man alone the power of producing value. A more correct analysis shows us that value is owing
to the action of labour, or rather the industry of man, combined
with the action of those agents which nature supplies, and with
that of capital. His ignorance of this principle prevented him
from establishing the true theory of the influence of machinery
in the production of riches."

In contradiction to the opinion of Adam Smith, M. Say, in the
fourth chapter, speaks of the value which is given to commodities by natural agents, such as the sun, the air, the pressure
of the atmosphere, etc., which are sometimes substituted for
the labour of man, and sometimes concur with him in producing.[1]
But these natural agents, though they add greatly to *value in use*,
never add exchangeable value, of which M. Say is speaking, to
a commodity: as soon as by the aid of machinery, or by the
knowledge of natural philosophy, you oblige natural agents to
do the work which was before done by man, the exchangeable
value of such work falls accordingly. If ten men turned a corn
mill, and it be discovered that by the assistance of wind, or of
water, the labour of these ten men may be spared, the flour
which is the produce partly of the work performed by the mill,

---

[1] " The first man who knew how to soften metals by fire is not the
creator of the value which that process adds to the melted metal. That
value is the result of the physical action of fire added to the industry and
capital of those who availed themselves of this knowledge."
" From this error Smith has drawn this false result, that the value of all
productions represents the recent or former labour of man, *or, in other
words, that riches are nothing else but accumulated labour ; from which, by a
second consequence, equally false, labour is the sole measure of riches, or of
the value of productions*."—Chap. iv. p. 31. The inferences with which
M. Say concludes are his own and not Dr. Smith's; they are correct if
no distinction be made between value and riches, and in this passage
M. Say makes none: but though Adam Smith, who defined riches to
consist in the abundance of necessaries, convenience, and enjoyments of
human life, would have allowed that machines and natural agents might
very greatly add to the riches of a country, he would not have allowed that
they add anything to the value of those riches.

would immediately fall in value, in proportion to the quantity of labour saved; and the society would be richer by the commodities which the labour of the ten men could produce, the funds destined for their maintenance being in no degree impaired. M. Say constantly overlooks the essential difference that there is between value in use and value in exchange.

M. Say accuses Dr. Smith of having overlooked the value which is given to commodities by natural agents, and by machinery, because he considered that the value of all things was derived from the labour of man; but it does not appear to me that this charge is made out; for Adam Smith nowhere undervalues the services which these natural agents and machinery perform for us, but he very justly distinguishes the nature of the value which they add to commodities—they are serviceable to us, by increasing the abundance of productions, by making men richer, by adding to value in use; but as they perform their work gratuitously, as nothing is paid for the use of air, of heat, and of water, the assistance which they afford us adds nothing to value in exchange.

CHAPTER XXI

FROM the account which has been given of the profits of stock, it will appear that no accumulation of capital will permanently lower profits unless there be some permanent cause for the rise of wages. If the funds for the maintenance of labour were doubled, trebled, or quadrupled, there would not long be any difficulty in procuring the requisite number of hands to be employed by those funds; but owing to the increasing difficulty of making constant additions to the food of the country, funds of the same value would probably not maintain the same quantity of labour. If the necessaries of the workman could be constantly increased with the same facility, there could be no permanent alteration in the rate of profit or wages, to whatever amount capital might be accumulated. Adam Smith, however, uniformly ascribes the fall of profits to the accumulation of capital, and to the competition which will result from it, without ever adverting to the increasing difficulty of providing food for the additional number of labourers which the additional capital will employ. " The increase of stock," he says, " which raises wages, tends to lower profit. When the stocks of many rich merchants are turned into the same trade, their mutual competition naturally tends to lower its profit; and when there is a like increase of stock in all the different trades carried on in the same society, the same competition must produce the same effect in all." Adam Smith speaks here of a rise of wages, but it is of a temporary rise, proceeding from increased funds before the population is increased; and he does not appear to see that at the same time that capital is increased the work to be effected by capital is increased in the same proportion. M. Say has, however, most satisfactorily shown that there is no amount of capital which may not be employed in a country, because a demand is only limited by production. No man produces but with a view to consume or sell, and he never sells but with an intention to purchase some other commodity, which may be immediately useful to him, or which may contribute to future production. By producing, then, he necessarily becomes either the consumer of his own goods, or the purchaser and consumer of the goods of some other person. It is not to be supposed that he should, for any length of time, be ill-informed of the commodities which he can most advantageously produce, to attain the object which he has in view, namely, the possession of other goods; and, therefore, it is not probable that he will continually produce a commodity for which there is no demand.[1]

There cannot, then, be accumulated in a country any amount of capital which cannot be employed productively until wages rise so high in consequence of the rise of necessaries, and so little consequently remains for the profits of stock, that the motive for accumulation ceases.[2] While the profits of stock are high, men will have a motive to accumulate. Whilst a man has any wished-for gratification unsupplied, he will have a demand for more commodities; and it will be an effectual demand while he has any new value to offer in exchange for them. If ten thousand pounds were given to a man having £100,000 per annum, he would not lock it up in a chest, but would either increase his expenses by £10,000, employ it himself productively, or lend it to some other person for that purpose; in either case, demand would be increased, although it would be for different objects. If he increased his expenses, his effectual demand might probably be for buildings, furniture, or some such enjoyment. If he employed his £10,000 productively, his effectual demand would be for food, clothing, and raw material, which might set new labourers to work; but still it would be demand.[3]

[1] Adam Smith speaks of Holland as affording an instance of the fall of profits from the accumulation of capital, and from every employment being consequently overcharged. "The government there borrow at 2 per cent., and private people of good credit at 3 per cent." But it should be remembered that Holland was obliged to import almost all the corn which she consumed, and by imposing heavy taxes on the necessaries of the labourer she further raised the wages of labour. These facts will sufficiently account for the low rate of profits and interest in Holland.

[2] Is the following quite consistent with M. Say's principle? "The more disposable capitals are abundant in proportion to the extent of employment for them, the more will the rate of interest on loans of capital fall."— Vol. ii. p. 108. If capital to any extent can be employed by a country, how can it be said to be abundant, compared with the extent of employment for it?

[3] Adam Smith says that, "When the produce of any particular branch of industry exceeds what the demand of the country requires, the surplus must be sent abroad, and exchanged for something for which there is a demand at home. *Without such exportation, a part of the productive labour of the country must cease, and the value of its annual produce diminish.* The land and labour of Great Britain produce generally more corn, woollens, and hardware than the demand of the home market requires. The surplus part of them, therefore, must be sent abroad, and exchanged for something for which there is a demand at home. It is only by means of such exporta-

Productions are always bought by productions, or by services; money is only the medium by which the exchange is effected. Too much of a particular commodity may be produced, of which there may be such a glut in the market as not to repay the capital expended on it; but this cannot be the case with respect to all commodities; the demand for corn is limited by the mouths which are to eat it, for shoes and coats by the persons who are to wear them; but though a community, or a part of a community, may have as much corn, and as many hats and shoes as it is able, or may wish to consume, the same cannot be said of every commodity produced by nature or by art. Some would consume more wine if they had the ability to procure it. Others, having enough of wine, would wish to increase the quantity or improve the quality of their furniture. Others might wish to ornament their grounds, or to enlarge their houses. The wish to do all or some of these is implanted in every man's breast; nothing is required but the means, and nothing can afford the means but an increase of production. If I had food and necessaries at my disposal, I should not be long in want of workmen who would put me in possession of some of the objects most useful or most desirable to me.

Whether these increased productions and the consequent demand which they occasion shall or shall not lower profits, depends solely on the rise of wages; and the rise of wages, excepting for a limited period, on the facility of producing the food and necessaries of the labourer. I say excepting for a limited period, because no point is better established, than that the supply of labourers will always ultimately be in proportion to the means of supporting them.

There is only one case, and that will be temporary, in which the accumulation of capital with a low price of food may be attended with a fall of profits; and that is when the funds for the maintenance of labour increase much more rapidly than

tion that this surplus can acquire a value sufficient to compensate the labour and expense of producing it." One would be led to think by the above passage that Adam Smith concluded we were under some necessity of producing a surplus of corn, woollen goods, and hardware, and that the capital which produced them could not be otherwise employed. It is, however, always a matter of choice in what way a capital shall be employed, and therefore there can never for any length of time be a surplus of any commodity; for if there were, it would fall below its natural price, and capital would be removed to some more profitable employment. No writer has more satisfactorily and ably shown than Dr. Smith the tendency of capital to move from employments in which the goods produced do not repay by their price the whole expenses, including the ordinary profits, of producing and bringing them to market.—See chap. x. book i.

population;—wages will then be high and profits low. If every man were to forego the use of luxuries, and be intent only on accumulation, a quantity of necessaries might be produced for which there could not be any immediate consumption. Of commodities so limited in number there might undoubtedly be a universal glut, and consequently there might neither be demand for an additional quantity of such commodities nor profits on the employment of more capital. If men ceased to consume, they would cease to produce. This admission does not impugn the general principle. In such a country as England, for example, it is difficult to suppose that there can be any disposition to devote the whole capital and labour of the country to the production of necessaries only.

When merchants engage their capitals in foreign trade, or in the carrying trade, it is always from choice and never from necessity: it is because in that trade their profits will be somewhat greater than in the home trade.

Adam Smith has justly observed " that the desire of food is limited in every man by the narrow capacity of the human stomach, but the desire of the conveniences and ornaments of building, dress, equipage, and household furniture seems to have no limit or certain boundary." Nature, then, has neces-sarily limited the amount of capital which can at any one time be profitably engaged in agriculture, but she has placed no limits to the amount of capital that may be employed in pro-curing " the conveniences and ornaments " of life. To procure these gratifications in the greatest abundance is the object in view, and it is only because foreign trade, or the carrying trade, will accomplish it better, that men engage in them in preference to manufacturing the commodities required, or a substitute for them, at home. If, however, from peculiar circumstances, we were precluded from engaging capital in foreign trade, or in the carrying trade, we should, though with less advantage, employ it at home; and while there is no limit to the desire of " con-veniences, ornaments of building, dress, equipage, and house-hold furniture," there can be no limit to the capital that may be employed in procuring them, except that which bounds our power to maintain the workmen who are to produce them.

Adam Smith, however, speaks of the carrying trade as one not of choice, but of necessity; as if the capital engaged in it would be inert if not so employed, as if the capital in the home trade could overflow if not confined to a limited amount. He says, " when the capital stock of any country is increased to such a degree *that it cannot be all employed in supplying the consumption, and supporting the productive labour of that particular country,* the surplus part of it naturally disgorges itself into the carrying trade, and is employed in performing the same offices to other countries."

"About ninety-six thousand hogheads of tobacco are annually purchased with a part of the surplus produce of British industry. But the demand of Great Britain does not require, perhaps, more than fourteen thousand. If the remaining eighty-two thousand, therefore, could not be sent abroad *and exchanged for something more in demand at home,* the importation of them would cease immediately, *and with it the productive labour of all the inhabitants of Great Britain who are at present employed in preparing the goods with which these eighty-two thousand hogsheads are annually purchased.*" But could not this portion of the productive labour of Great Britain be employed in preparing some other sort of goods, with which something more in demand at home might be purchased? And if it could not, might we not employ this productive labour, though with less advantage, in making those goods in demand at home, or at least some substitute for them? If we wanted velvets, might we not attempt to make velvets; and if we could not succeed, might we not make more cloth, or some other object desirable to us?

We manufacture commodities, and with them buy goods abroad, because we can obtain a greater quantity than we could make at home. Deprive us of this trade, and we immediately manufacture again for ourselves. But this opinion of Adam Smith is at variance with all his general doctrines on this subject. " If a foreign country can supply us with a commodity cheaper than we ourselves can make it, better buy it of them with some part of the produce of our own industry, employed in a way in which we have some advantage. *The general industry of the country, being always in proportion to the capital which employs it,* will not thereby be diminished, but only left to find out the way in which it can be employed with the greatest advantage."

Again. " Those, therefore, who have the command of more food than they themselves can consume, are always willing to exchange the surplus, or, what is the same thing, the price of it, for gratifications of another kind. What is over and above satisfying the limited desire is given for the amusement of those desires which cannot be satisfied, but seem to be altogether endless. The poor, in order to obtain food, exert themselves to gratify those fancies of the rich; and to obtain it more certainly, they vie with one another in the cheapness and perfection of their work. The number of workmen increases with the increasing quantity of food, or with the growing improvement and cultivation of the lands; and as the nature of their business admits of the utmost subdivisions of labours, the quantity of materials which they can work up increases in a much greater proportion than their numbers. Hence arises a demand for every sort of material which human invention can employ, either usefully or ornamentally, in building, dress, equipage, or household furniture; for the fossils and minerals contained in the bowels of the earth, the precious metals, and the precious stones."

It follows, then, from these admissions, that there is no limit to demand—no limit to the employment of capital while it yields any profit, and that, however abundant capital may become, there is no other adequate reason for a fall of profit but a rise of wages, and further, it may be added that the only adequate and permanent cause for the rise of wages is the increasing difficulty of providing food and necessaries for the increasing number of workmen.

Adam Smith has justly observed that it is extremely difficult to determine the rate of the profits of stock. " Profit is so fluctuating that even in a particular trade, and much more in trades in general, it would be difficult to state the average rate of it. To judge of what it may have been formerly, or in remote periods of time, with any degree of precision, must be altogether impossible." Yet since it is evident that much will be given for the use of money when much can be made by it, he suggests that " the market rate of interest will lead us to form some notion of the rate of profits, and the history of the progress of interest afford us that of the progress of profits." Undoubtedly, if the market rate of interest could be accurately known for any considerable period, we should have a tolerably correct criterion by which to estimate the progress of profits.

But in all countries, from mistaken notions of policy, the state has interfered to prevent a fair and free market rate of interest by imposing heavy and ruinous penalties on all those who shall take more than the rate fixed by law. In all countries probably these laws are evaded, but records give us little information on this head, and point out rather the legal and fixed rate than the market rate of interest. During the present war,

Exchequer and Navy Bills have been frequently at so high a discount as to afford the purchasers of them 7, 8 per cent., or a greater rate of interest for their money. Loans have been raised by government at an interest exceeding 6 per cent., and individuals have been frequently obliged, by indirect means, to pay more than 10 per cent. for the interest of money; yet during this same period the legal rate of interest has been uniformly at 5 per cent. Little dependence for information, then, can be placed on that which is the fixed and legal rate of interest, when we find it may differ so considerably from the market rate. Adam Smith informs us that from the 37th of Henry VIII. to 21st of James I., 10 per cent. continued to be the legal rate of interest. Soon after the restoration, it was reduced to 6 per cent., and by the 12th of Anne to 5 per cent. He thinks the legal rate followed, and did not precede, the market rate of interest. Before the American war, government borrowed at 3 per cent., and the people of credit in the capital and in many other parts of the kingdom at $3\frac{1}{2}$, 4, and $4\frac{1}{2}$ per cent.

The rate of interest, though ultimately and permanently governed by the rate of profit, is, however, subject to temporary variations from other causes. With every fluctuation in the quantity and value of money, the prices of commodities naturally vary. They vary also, as we have already shown, from the alteration in the proportion of supply to demand, although there should not be either greater facility or difficulty of production. When the market prices of goods fall from an abundant supply, from a diminished demand, or from a rise in the value of money, a manufacturer naturally accumulates an unusual quantity of finished goods, being unwilling to sell them at very depressed prices. To meet his ordinary payments, for which he used to depend on the sale of his goods, he now endeavours to borrow on credit, and is often obliged to give an increased rate of interest. This, however, is but of temporary duration; for either the manufacturer's expectations were well grounded, and the market price of his commodities rises, or he discovers that there is a permanently diminished demand, and he no longer resists the course of affairs: prices fall, and money and interest regain their real value. If, by the discovery of a new mine, by the abuses of banking, or by any other cause, the quantity of money be greatly increased, its ultimate effect is to raise the prices of commodities in proportion to the increased quantity of money; but there is probably always an interval during which some effect is produced on the rate of interest.

The price of funded property is not a steady criterion by which to judge of the rate of interest. In time of war, the stock market is so loaded by the continual loans of government that the price of stock has not time to settle at its fair level before a new operation of funding takes place, or it is affected by anticipation of political events. In time of peace, on the contrary, the operations of the sinking fund, the unwillingness which a particular class of persons feel to divert their funds to any other employment than that to which they have been accustomed, which they think secure, and in which their dividends are paid with the utmost regularity, elevates the price of stock, and consequently depresses the rate of interest on these securities below the general market rate. It is observable, too, that for different securities government pays very different rates of interest. Whilst £100 capital in 5 per cent. stock is selling for £95, an exchequer bill of £100 will be sometimes selling for £100 5s., for which exchequer bill no more interest will be annually paid than £4 11s. 3d.: one of these securities pays to a purchaser, at the above prices, an interest of more than 5¼ per cent., the other but little more than 4¼; a certain quantity of these exchequer bills is required as a safe and marketable investment for bankers; if they were increased much beyond this demand they would probably be as much depreciated as the 5 per cent. stock. A stock paying 3 per cent. per annum will always sell at a proportionally greater price than stock paying 5 per cent., for the capital debt of neither can be discharged but at par, or £100 money for £100 stock. The market rate of interest may fall to 4 per cent., and government would then pay the holder of 5 per cent. stock at par, unless he consented to take 4 per cent. on some diminished rate of interest under 5 per cent.: they would have no advantage from so paying the holder of 3 per cent. stock till the market rate of interest had fallen below 3 per cent. per annum. To pay the interest on the national debt large sums of money are withdrawn from circulation four times in the year for a few days. These demands for money being only temporary seldom affect prices; they are generally surmounted by the payment of a large rate of interest.[1]

---

[1] " All kinds of public loans," observes M. Say, " are attended with the inconvenience of withdrawing capital, or portions of capital, from productive employments, to devote them to consumption; and when they take place in a country, *the government of which does not inspire much confidence*, they have the further inconvenience of raising the interest of capital. Who would lend at 5 per cent. per annum to agriculture, to

manufacturers, and to commerce, when a borrower may be found ready to pay an interest of 7 or 8 per cent.? That sort of income which is called profit of stock would rise then at the expense of the consumer. Consumption would be reduced by the rise in the price of produce; and the other productive services would be less in demand, less well paid. The whole nation, capitalists excepted, would be the sufferers from such a state of things." To the question, " who would lend money to farmers, manufacturers, and merchants, at 5 per cent. per annum, when another borrower, having little credit, would give 7 or 8? " I reply, that every prudent and reasonable man would. Because the rate of interest is 7 or 8 per cent. there where the lender runs extraordinary risk is this any reason that it should be equally high in those places where they are secured from such risks? M. Say allows that the rate of interest depends on the rate of profits; but it does not therefore follow that the rate of profits depends on the rate of interest. One is the cause, the other the effect, and it is impossible for any circumstances to make them change places.

CHAPTER XXII

BOUNTIES ON EXPORTATION, AND PROHIBITIONS OF IMPORTATION

A BOUNTY on the exportation of corn tends to lower its price to the foreign consumer, but it has no permanent effect on its price in the home market.

Suppose that to afford the usual and general profits of stock, the price of corn should in England be £4 per quarter; it could not then be exported to foreign countries where it sold for £3 15s. per quarter. But if a bounty of 10s. per quarter were given on exportation, it could be sold in the foreign market at £3 10s., and consequently the same profit would be afforded to the corn grower whether he sold it at £3 10s. in the foreign or at £4 in the home market.

A bounty then, which should lower the price of British corn in the foreign country below the cost of producing corn in that country, would naturally extend the demand for British and diminish the demand for their own corn. This extension of demand for British corn could not fail to raise its price for a time in the home market, and during that time to prevent also its falling so low in the foreign market as the bounty has a tendency to effect. But the causes which would thus operate on the market price of corn in England would produce no effect whatever on its natural price, or its real cost of production. To grow corn would neither require more labour nor more capital, and, consequently, if the profits of the farmer's stock were before only equal to the profits of the stock of other traders, they will, after the rise of price, be considerably above them. By raising the profits of the farmer's stock, the bounty will operate as an encouragement to agriculture, and capital will be withdrawn from manufactures to be employed on the land till the enlarged demand for the foreign market has been supplied, when the price of corn will again fall in the home market to its natural and necessary price, and profits will be again at their ordinary and accustomed level. The increased supply of grain operating on the foreign market will also lower its price in the country to which it is exported, and will thereby restrict the profits of the exporter to the lowest rate at which he can afford to trade.

The ultimate effect then of a bounty on the exportation of corn is not to raise or to lower the price in the home market, but to lower the price of corn to the foreign consumer—to the whole extent of the bounty, if the price of corn had not before been lower in the foreign than in the home market—and in a less degree if the price in the home had been above the price in the foreign market.

A writer in the fifth volume of the *Edinburgh Review*, on the subject of a bounty on the exportation of corn, has very clearly pointed out its effects on the foreign and home demand. He has also justly remarked that it would not fail to give encouragement to agriculture in the exporting country; but he appears to have imbibed the common error which has misled Dr. Smith, and, I believe, most other writers on this subject. He supposes, because the price of corn ultimately regulates wages, that therefore it will regulate the price of all other commodities. He says that the bounty, " by raising the profits of farming, will operate as an encouragement to husbandry; by raising the price of corn to the consumers at home it will diminish for the time their power of purchasing this necessary of life, and thus abridge their real wealth. It is evident, however, that this last effect must be temporary: the wages of the labouring consumers had been adjusted before by competition, and the same principle will adjust them again to the same rate, by raising the money price of labour, *and through that, of other commodities, to the money price of corn*. The bounty upon exportation, therefore, will ultimately raise the money price of corn in the home market; not directly, however, but through the medium of an extended demand in the foreign market, and a consequent enhancement of the real price at home: *and this rise of the money price, when it has once been communicated to other commodities, will of course become fixed.*"

If, however, I have succeeded in showing that it is not the rise in the money wages of labour which raises the price of commodities, but that such rise always affects profits, it will follow that the prices of commodities would not rise in consequence of a bounty.

But a temporary rise in the price of corn, produced by an increased demand from abroad, would have no effect on the money price of labour. The rise of corn is occasioned by a competition for that supply which was before exclusively appropriated to the home market. By raising profits, additional capital is employed in agriculture, and the increased supply is obtained; but till it be obtained, the high price is absolutely necessary to proportion the consumption to the supply, which would be counteracted by a rise of wages. The rise of corn is the consequence of its scarcity, and is the means by which the demand of the home purchasers is diminished. If wages were increased, the competition would increase, and a further rise of the price of corn would become necessary. In this account of the effects of a bounty nothing has been supposed to occur to raise the natural price of corn, by which its market price is ultimately governed; for it has not been supposed that any additional labour would be required on the land to insure a given production, and this alone can raise its natural price. If the natural price of cloth were 20s. per yard, a great increase in the foreign demand might raise the price to 25s., or more, but the profits which would then be made by the clothier would not fail to attract capital in that direction, and although the demand should be doubled, trebled, or quadrupled, the supply would ultimately be obtained, and cloth would fall to its natural price of 20s. So, in the supply of corn, although we should export 200,000, 300,000, or 800,000 quarters annually, it would ultimately be produced at its natural price, which never varies, unless a different quantity of labour becomes necessary to production.

Perhaps in no part of Adam Smith's justly celebrated work are his conclusions more liable to objection than in the chapter on bounties. In the first place, he speaks of corn as of a commodity of which the production cannot be increased in consequence of a bounty on exportation; he supposes invariably that it acts only on the quantity actually produced, and is no stimulus to farther production. "In years of plenty," he says, "by occasioning an extraordinary exportation, it necessarily keeps up the price of corn in the home market above what it would naturally fall to. In years of scarcity, though the bounty is frequently suspended, yet the great exportation which it occasions in years of plenty must frequently hinder, more or less, the plenty of one year from relieving the scarcity of another. Both in the years of plenty and in years of scarcity, therefore, the bounty necessarily tends to raise the money price of corn somewhat higher than it otherwise would be in the home market." [1]

---

[1] In another place he says, that " whatever extension of the foreign market can be occasioned by the bounty must, in every particular year, be altogether at the expense of the home market, as every bushel of corn which

Adam Smith appears to have been fully aware that the correctness of his argument entirely depended on the fact whether the increase " of the money price of corn, by rendering that commodity more profitable to the farmer, would not necessarily encourage its production."

" I answer," he says, " that this might be the case if the effect of the bounty was to raise the real price of corn, or to enable the farmer, with an equal quantity of it, to maintain a greater number of labourers in the same manner, whether liberal, moderate, or scanty, as other labourers are commonly maintained in his neighbourhood."

If nothing were consumed by the labourer but corn, and if the portion which he received was the very lowest which his sustenance required, there might be some ground for supposing that the quantity paid to the labourer could, under no circumstances, be reduced—but the money wages of labour sometimes do not rise at all, and never rise in proportion to the rise in the money price of corn, because corn, though an important part, is only a part of the consumption of the labourer. If half his wages were expended on corn, and the other half on soap, candles, fuel, tea, sugar, clothing, etc., commodities on which no rise is supposed to take place, it is evident that he would be quite as well paid with a bushel and a half of wheat when it was 16s. a bushel, as he was with two bushels when the price was 8s. per bushel; or with 24s. in money as he was before with 16s. His wages would rise only 50 per cent. though corn rose 100 per cent.; and, consequently, there would be sufficient motive to divert more capital to the land if profits on other

is exported by means of the bounty, and which would not have been exported without the bounty, would have remained in the home market to increase the consumption and to lower the price of that commodity. The corn bounty, it is to be observed, as well as every other bounty upon exportation, imposes two different taxes upon the people:—first, the tax which they are obliged to contribute in order to pay the bounty; and, secondly, the tax which arises from the advanced price of the commodity in the home market, and which, as the whole body of the people are purchasers of corn, must, in this particular commodity, be paid by the whole body of the people. In this particular commodity, therefore, this second tax is by much the heaviest of the two." " For every five shillings, therefore, which they contribute to the payment of the first tax, they must contribute six pounds four shillings to the payment of the second." " The extraordinary exportation of corn, therefore, occasioned by the bounty, not only in every particular year diminishes the home just as much as it extends the foreign market and consumption; but, by restraining the population and industry of the country, its final tendency is to stunt and restrain the gradual extension of the home market, and thereby, in the long run, rather to diminish than to augment the whole market and consumption of corn."

trades continued the same as before. But such a rise of wages
would also induce manufacturers to withdraw their capitals
from manufactures to employ them on the land; for, whilst the
farmer increased the price of his commodity 100 per cent. and
his wages only 50 per cent., the manufacturer would be obliged
also to raise wages 50 per cent., whilst he had no compensation
whatever in the rise of his manufactured commodity for this
increased charge of production; capital would consequently
flow from manufactures to agriculture, till the supply would
again lower the price of corn to 8s. per bushel and wages to
16s. per week; when the manufacturer would obtain the same
profits as the farmer, and the tide of capital would cease to set
in either direction. This is, in fact, the mode in which the
cultivation of corn is always extended, and the increased wants
of the market supplied. The funds for the maintenance of
labour increase, and wages are raised. The comfortable situation
of the labourer induces him to marry—population increases, and
the demand for corn raises its price relatively to other things—
more capital is profitably employed on agriculture, and continues
to flow towards it, till the supply is equal to the demand, when
the price again falls, and agricultural and manufacturing profits
are again brought to a level.

But whether wages were stationary after the rise in the price
of corn, or advanced moderately or enormously, is of no importance to this question, for wages are paid by the manufacturer
as well as by the farmer, and, therefore, in this respect they
must be equally affected by a rise in the price of corn. But
they are unequally affected in their profits, inasmuch as the
farmer sells his commodity at an advanced price, while the
manufacturer sells his for the same price as before. It is, however, the inequality of profit which is always the inducement
to remove capital from one employment to another; and,
therefore, more corn would be produced, and fewer commodities
manufactured. Manufactures would not rise, because fewer
would be manufactured, for a supply of them would be obtained
in exchange for the exported corn.

A bounty, if it raises the price of corn, either raises it in comparison with the price of other commodities or it does not. If
the affirmative be true, it is impossible to deny the greater
profits of the farmer, and the temptation to the removal of
capital till its price is again lowered by an abundant supply.
If it does not raise it in comparison with other commodities,
where is the injury to the home consumer beyond the incon-

venience of paying the tax? If the manufacturer pays a greater price for his corn, he is compensated by the greater price at which he sells his commodity, with which his corn is ultimately purchased.

The error of Adam Smith proceeds precisely from the same source as that of the writer in the *Edinburgh Review;* for they both think " that the money price of corn regulates that of all other home-made commodities." [1] " It regulates," says Adam Smith, " the money price of labour, which must always be such as to enable the labourer to purchase a quantity of corn sufficient to maintain him and his family, either in the liberal, moderate, or scanty manner, in which the advancing, stationary, or declining circumstances of the society oblige his employers to maintain him. By regulating the money price of all the other parts of the rude produce of land, it regulates that of the materials of almost all manufactures. By regulating the money price of labour, it regulates that of manufacturing art and industry; and by regulating both, it regulates that of the complete manufacture. *The money price of labour, and of everything that is the produce either of land or labour, must necessarily rise or fall in proportion to the money price of corn.*"

This opinion of Adam Smith I have before attempted to refute. In considering a rise in the price of commodities as a necessary consequence of a rise in the price of corn, he reasons as though there were no other fund from which the increased charge could be paid. He has wholly neglected the consideration of profits, the diminution of which forms that fund, without raising the price of commodities. If this opinion of Dr. Smith were well founded, profits could never really fall, whatever accumulation of capital there might be. If, when wages rose, the farmer could raise the price of his corn, and the clothier, the hatter, the shoemaker, and every other manufacturer could also raise the price of their goods in proportion to the advance, although estimated in money they might be all raised, they would continue to bear the same value relatively to each other. Each of these trades could command the same quantity as before of the goods of the others, which, since it is goods, and not money, which constitute wealth, is the only circumstance that could be of importance to them; and the whole rise in the price of raw produce and of goods would be injurious to no other persons but to those whose property consisted of gold and silver, or whose annual income was paid in a contributed quantity

[1] The same opinion is held by M. Say.—Vol. ii. p. 335.

of those metals, whether in the form of bullion or of money. Suppose the use of money to be wholly laid aside, and all trade to be carried on by barter. Under such circumstances, could corn rise in exchangeable value with other things? If it could, then it is not true that the value of corn regulates the value of all other commodities; for to do that, it should not vary in relative value to them. If it could not, then it must be maintained that whether corn be obtained on rich or on poor land, with much labour or with little, with the aid of machinery or without, it would always exchange for an equal quantity of all other commodities.

I cannot, however, but remark that though Adam Smith's general doctrines correspond with this which I have just quoted, yet in one part of his work he appears to have given a correct account of the nature of value. " The proportion between the value of gold and silver, and that of goods of any other kind, DEPENDS IN ALL CASES," he says, " *upon the proportion between the quantity of labour which is necessary in order to bring a certain quantity of gold and silver to market, and that which is necessary to bring thither a certain quantity of any other sort of goods.*" Does he not here fully acknowledge, that if any increase takes place in the quantity of labour required to bring one sort of goods to market, whilst no such increase takes place in bringing another sort thither, the first sort will rise in relative value? If no more labour than before be required to bring either cloth or gold to market, they will not vary in relative value, but if more labour be required to bring corn and shoes to market, will not corn and shoes rise in value relatively to cloth and money made of gold?

Adam Smith again considers that the effect of the bounty is to cause a partial degradation in the value of money. " That degradation," says he, " in the value of silver which is the effect of the fertility of the mines, and which operates equally, or very nearly equally, through the greater part of the commercial world, is a matter of very little consequence to any particular country. The consequent rise of all money prices, though it does not make those who receive them really richer, does not make them really poorer. A service of plate becomes really cheaper, and everything else remains precisely of the same real value as before." This observation is most correct.

" But that degradation in the value of silver, which, being the effect either of the peculiar situation or of the political institutions of a particular country, takes place only in that country, is a matter of very great consequence, which, far from tending to make anybody really richer, tends to make everybody really poorer. The rise in the money price of all commodities, which is in this case peculiar to that country, tends to discourage more or less every sort of industry which is carried on within it, and to enable foreign nations, by furnishing almost all sorts of goods for a smaller quantity of silver than its own workmen can afford to do, to undersell them, not only in the foreign, but even in the home market."

I have elsewhere attempted to show that a partial degradation in the value of money, which shall affect both agricultural produce and manufactured commodities, cannot possibly be permanent. To say that money is partially degraded, in this sense, is to say that all commodities are at a high price; but while gold and silver are at liberty to make purchases in the cheapest market, they will be exported for the cheaper goods of other countries, and the reduction of their quantity will increase their value at home; commodities will regain their usual level, and those fitted for foreign markets will be exported as before.

A bounty, therefore, cannot, I think, be objected to on this ground.

If, then, a bounty raises the price of corn in comparison with all other things, the farmer will be benefited, and more land will be cultivated; but if the bounty do not raise the value of corn relatively to other things then no other inconvenience will attend it than that of paying the bounty; one which I neither wish to conceal nor underrate.

Dr. Smith states that " by establishing high duties on the importation, and bounties on the exportation of corn, the country gentlemen seemed to have imitated the conduct of the manufacturers." By the same means, both had endeavoured to raise the value of their commodities. " They did not, perhaps, attend to the great and essential difference which nature has established between corn and almost every other sort of goods. When by either of the above means you enable our manufacturers to sell their goods for somewhat a better price than they otherwise could get for them, you raise not only the nominal, but the real price of those goods. You increase not only the nominal, but the real profit, the real wealth and revenue of those manufacturers — you really encourage those manufacturers. But when, by the like institutions, you raise the nominal or money price of corn, you do not raise its real value, you do not increase the real wealth of our farmers or country gentlemen, you do not encourage the growth of corn. The nature of things has stamped upon corn a real value which cannot be altered by merely altering its money price. Through the world in general that value is equal to the quantity of labour which it can maintain."

I have already attempted to show that the market price of corn would, under an increased demand from the effects of a bounty, exceed its natural price, till the requisite additional supply was obtained, and that then it would again fall to its natural price. But the natural price of corn is not so fixed as the natural price of commodities; because, with any great additional demand for corn, land of a worse quality must be taken into cultivation, on which more labour will be required to produce a given quantity, and the natural price of corn will be raised. By a continued bounty, therefore, on the exportation of corn, there would be created a tendency to a permanent rise in the price of corn, and this, as I have shown elsewhere,[1] never fails to raise rent. Country gentlemen, then, have not only a temporary but a permanent interest in prohibitions of the importation of corn, and in bounties on its exportation; but manufacturers have no permanent interest in establishing high duties on the importation, and bounties on the exportation of commodities; their interest is wholly temporary.

A bounty on the exportation of manufactures will, undoubtedly, as Dr. Smith contends, raise for a time the market price of manufactures, but it will not raise their natural price. The labour of 200 men will produce double the quantity of these goods that 100 could produce before; and, consequently, when the requisite quantity of capital was employed in supplying the requisite quantity of manufactures, they would again fall to their natural price, and all advantage from a high market price would cease. It is, then, only during the interval after the rise in the market price of commodities, and till the additional supply is obtained, that the manufacturers will enjoy high profits; for as soon as prices had subsided, their profits would sink to the general level.

Instead of agreeing, therefore, with Adam Smith, that the country gentlemen had not so great an interest in prohibiting the importation of corn, as the manufacturer had in prohibiting the importation of manufactured goods, I contend, that they have a much superior interest; for their advantage is permanent,

[1] See chapter on Rent.

while that of the manufacturer is only temporary. Dr. Smith observes that nature has established a great and essential difference between corn and other goods, but the proper inference from that circumstance is directly the reverse of that which he draws from it; for it is on account of this difference that rent is created, and that country gentlemen have an interest in the rise of the natural price of corn. Instead of comparing the interest of the manufacturer with the interest of the country gentleman, Dr. Smith should have compared it with the interest of the farmer, which is very distinct from that of his landlord. Manufacturers have no interest in the rise of the natural price of their commodities, nor have farmers any interest in the rise of the natural price of corn, or other raw produce, though both these classes are benefited while the market price of their productions exceeds their natural price. On the contrary, landlords have a most decided interest in the rise of the natural price of corn; for the rise of rent is the inevitable consequence of the difficulty of producing raw produce, without which its natural price could not rise. Now, as bounties on exportation and prohibitions of the importation of corn increase the demand, and drive us to the cultivation of poorer lands, they necessarily occasion an increased difficulty of production.

The sole effect of high duties on the importation, either of manufactures or of corn, or of a bounty on their exportation, is to divert a portion of capital to an employment which it would not naturally seek. It causes a pernicious distribution of the general funds of the society—it bribes a manufacturer to commence or continue in a comparatively less profitable employment. It is the worst species of taxation, for it does not give to the foreign country all that it takes away from the home country, the balance of loss being made up by the less advantageous distribution of the general capital. Thus, if the price of corn is in England £4, and in France £3 15s., a bounty of 10s. will ultimately reduce it to £3 10s. in France, and maintain it at the same price of £4 in England. For every quarter exported, England pays a tax of 10s. For every quarter imported into France, France gains only 5s., so that the value of 5s. per quarter is absolutely lost to the world by such a distribution of its funds, as to cause diminished production, probably not of corn, but of some other object of necessity or enjoyment.

Mr. Buchanan appears to have seen the fallacy of Dr. Smith's arguments respecting bounties, and on the last passage which I have quoted very judiciously remarks: " In asserting that nature has stamped a real value on corn, which cannot be altered by merely altering its money price, Dr. Smith confounds its value in use with its value in exchange. A bushel of wheat will not feed more people during scarcity than during plenty; but a bushel of wheat will exchange for a greater quantity of luxuries and conveniences when it is scarce than when it is abundant; and the landed proprietors, who have a surplus of food to dispose of, will therefore, in times of scarcity, be richer men; they will exchange their surplus for a greater value of other enjoyments than when corn is in greater plenty. It is vain to argue, therefore, that if the bounty occasions a forced exportation of corn, it will not also occasion a real rise of price." The whole of Mr. Buchanan's arguments on this part of the subject of bounties appear to me to be perfectly clear and satisfactory.

Mr. Buchanan, however, has not, I think, any more than Dr. Smith or the writer in the *Edinburgh Review,* correct opinions as to the influence of a rise in the price of labour on manufactured commodities. From his peculiar views, which I have elsewhere noticed, he thinks that the price of labour has no connection with the price of corn, and, therefore, that the real value of corn might and would rise without affecting the price of labour; but if labour were affected, he would maintain with Adam Smith and the writer in the *Edinburgh Review* that the price of manufactured commodities would also rise; and then I do not see how he would distinguish such a rise of corn from a fall in the value of money, or how he could come to any other conclusion than that of Dr. Smith. In a note to page 276, vol. i. of the *Wealth of Nations,* Mr. Buchanan observes, " but the price of corn does not regulate the money price of all the other parts of the rude produce of land. It regulates the price of neither metals, nor of various other useful substances, such as coals, wood, stones, etc.; *and as it does not regulate the price of labour, it does not regulate the price of manufactures ;* so that the bounty, in so far as it raises the price of corn, is undoubtedly a real benefit to the farmer. It is not on this ground, therefore, that its policy must be argued. Its encouragement to agriculture, by raising the price of corn, must be admitted; and the question then comes to be whether agriculture ought to be thus encouraged?" —It is then, according to Mr. Buchanan, a real benefit to the farmer, because it does not raise the price of labour; but if it did, it would raise the price of all things in proportion, and then it would afford no particular encouragement to agriculture.

It must, however, be conceded that the tendency of a bounty

on the exportation of any commodity is to lower in a small degree the value of money. Whatever facilitates exportation tends to accumulate money in a country; and, on the contrary, whatever impedes exportation tends to diminish it. The general effect of taxation, by raising the prices of the commodities taxed, tends to diminish exportation, and, therefore, to check the influx of money; and, on the same principle, a bounty encourages the influx of money. This is more fully explained in the general observations on taxation.

The injurious effects of the mercantile system have been fully exposed by Dr. Smith; the whole aim of that system was to raise the price of commodities in the home market by prohibiting foreign competition; but this system was no more injurious to the agricultural classes than to any other part of the community. By forcing capital into channels where it would not otherwise flow, it diminished the whole amount of commodities produced. The price, though permanently higher, was not sustained by scarcity, but by difficulty of production; and therefore, though the sellers of such commodities sold them for a higher price, they did not sell them, after the requisite quantity of capital was employed in producing them, at higher profits.[1]

The manufacturers themselves, as consumers, had to pay an additional price for such commodities, and, therefore, it cannot be correctly said that " the enhancement of price occasioned by both (corporation laws and high duties on the importations of foreign commodities) is everywhere finally paid by the landlords, farmers, and labourers of the country."

It is the more necessary to make this remark as in the present day the authority of Adam Smith is quoted by country gentlemen for imposing similar high duties on the importation of foreign corn. Because the cost of production, and, therefore,

---

[1] M. Say supposes the advantage of the manufacturers at home to be more than temporary. " A government which absolutely prohibits the importation of certain foreign goods establishes a monopoly *in favour of those* who produce such commodities at home *against those* who consume them; in other words, those at home who produce them having the exclusive privilege of selling them, may elevate their price above the natural price; and the consumers at home, not being able to obtain them elsewhere, are obliged to purchase them at a higher price."—Vol. i. p. 201. But how can they permanently support the market price of their goods above the natural price, when every one of their fellow citizens is free to enter into the trade? They are guaranteed against foreign, but not against home competition. The real evil arising to the country from such monopolies, if they can be called by that name, lies not in raising the market price of such goods, but in raising their real and natural price. By increasing the cost of production, a portion of the labour of the country is less productively employed.

the prices of various manufactured commodities, are raised to the consumer by one error in legislation, the country has been called upon, on the plea of justice, quietly to submit to fresh exactions. Because we all pay an additional price for our linen, muslin, and cottons, it is thought just that we should pay also an additional price for our corn. Because, in the general distribution of the labour of the world, we have prevented the greatest amount of productions from being obtained by our portion of that labour in manufactured commodities, we should further punish ourselves by diminishing the productive powers of the general labour in the supply of raw produce. It would be much wiser to acknowledge the errors which a mistaken policy has induced us to adopt, and immediately to commence a gradual recurrence to the sound principles of a universally free trade.[1]

" I have already had occasion to remark," observes M. Say, " in speaking of what is improperly called the balance of trade, that if it suits a merchant better to export the precious metals to a foreign country than any other goods, it is also the interest of the state that he should export them, because the state only gains or loses through the channel of its citizens; and in what concerns foreign trade, that which best suits the individual best suits also the state; therefore, by opposing obstacles to the exportation which individuals would be inclined to make of the precious metals, nothing more is done than to force them to substitute some other commodity less profitable to themselves and to the state. It must, however, be remarked that I say only *in what concerns foreign trade* ; because the profits which merchants make by their dealings with their countrymen, as well as those which are made in the exclusive commerce with colonies, are not entirely gains for the state. In the trade between individuals of the same country there is no other gain but the value of a utility produced; *que la valeur d'une utilité produite*," [2] vol. i. p. 401. I cannot see the distinction here

---

[1] " A freedom of trade is alone wanted to guarantee a country like Britain, abounding in all the varied products of industry, in merchandise suited to the wants of every society, from the possibility of a scarcity. The nations of the earth are not condemned to throw the dice to determine which of them shall submit to famine. There is always abundance of food in the world. To enjoy a constant plenty we have only to lay aside our prohibitions and restrictions, and cease to counteract the benevolent wisdom of Providence."—Article " Corn Laws and Trade," Supplement to *Encyclopædia Britannica.*

[2] Are not the following passages contradictory to the one above quoted? " Besides, that home trade, though less noticed (because it is in a variety of hands), is the most considerable, it is also the most profitable. The

made between the profits of the home and foreign trade. The object of all trade is to increase productions. If, for the purchase of a pipe of wine, I had it in my power to export bullion which was bought with the value of the produce of 100 days' labour, but government, by prohibiting the exportation of bullion, should oblige me to purchase my wine with a commodity bought with the value of the produce of 105 days' labour, the produce of five days' labour is lost to me, and, through me, to the state. But if these transactions took place between individuals in different provinces of the same country, the same advantage would accrue both to the individual, and, through him, to the country, if he were unfettered in his choice of the commodities with which he made his purchases, and the same disadvantage if he were obliged by government to purchase with the least beneficial commodity. If a manufacturer could work up with the same capital more iron where coals are plentiful than he could where coals are scarce, the country would be benefited by the difference. But if coals were nowhere plentiful, and he imported iron, and could get this additional quantity by the manufacture of a commodity with the same capital and labour, he would, in like manner, benefit his country by the additional quantity of iron. In the sixth chapter of this work I have endeavoured to show that all trade, whether foreign or domestic, is beneficial, by increasing the quantity and not by increasing the value of productions. We shall have no greater value whether we carry on the most beneficial home and foreign trade, or, in consequence of being fettered by prohibitory laws, we are obliged to content ourselves with the least advantageous. The rate of profits and the value produced will be the same. The advantage always resolves itself into that which M. Say appears to confine to the home trade; in both cases there is no other gain but that of the value of a *utilité produite*.

commodities exchanged in that trade are necessarily the productions of the same country."—Vol. i. p. 84.

" The English government has not observed that the most profitable sales are those which a country makes to itself, because they cannot take place without two values being produced by the nation; the value which is sold, and the value with which the purchase is made."—Vol. i. p. 221.

I shall, in the twenty-sixth chapter, examine the soundness of this opinion.

CHAPTER XXIII

ON BOUNTIES ON PRODUCTIONS

It may not be uninstructive to consider the effects of a bounty on the *production* of raw produce and other commodities, with a view to observe the application of the principles which I have been endeavouring to establish with regard to the profits of stock, the division of the annual produce of the land and labour, and the relative prices of manufactures and raw produce. In the first place, let us suppose that a tax was imposed on all commodities for the purpose of raising a fund to be employed by government in giving a bounty on the *production* of corn. As no part of such a tax would be expended by government, and as all that was received from one class of the people would be returned to another, the nation collectively would be neither richer nor poorer from such a tax and bounty. It would be readily allowed that the tax on commodities by which the fund was created would raise the price of the commodities taxed; all the consumers of those commodities, therefore, would contribute towards that fund; in other words, their natural or necessary price being raised, so would, too, their market price. But for the same reason that the natural price of those commodities would be raised, the natural price of corn would be lowered; before the bounty was paid on production, the farmers obtained as great a price for their corn as was necessary to repay them their rent and their expenses, and afford them the general rate of profits; after the bounty, they would receive more than that rate, unless the price of corn fell by a sum at least equal to the bounty. The effect, then, of the tax and bounty would be to raise the price of commodities in a degree equal to the tax levied on them, and to lower the price of corn by a sum equal to the bounty paid. It will be observed, too, that no permanent alteration could be made in the distribution of capital between agriculture and manufactures, because, as there would be no alteration either in the amount of capital or population, there would be precisely the same demand for bread and manufactures. The profits of the farmer would be no higher than the general level after the fall in the price of corn; nor would the profits of the manufacturer be lower after the rise of manufactured goods; the bounty, then, would not occasion any more capital to be employed on the land in the production of corn, nor any less in the manufacture of goods. But how would the interest of the landlord be affected? On the same principles that a tax on raw produce would lower the corn rent of land, leaving the money rent unaltered, a bounty on production, which is directly the contrary of a tax, would raise corn rent, leaving the money rent unaltered.[1] With the same money rent the landlord would have a greater price to pay for his manufactured goods, and a less price for his corn; he would probably, therefore, be neither richer nor poorer.

Now, whether such a measure would have any operation on the wages of labour would depend on the question whether the labourer, in purchasing commodities, would pay as much towards the tax as he would receive from the effects of the bounty in the low price of his food. If these two quantities were equal, wages would continue unaltered; but if the commodities taxed were not those consumed by the labourer, his wages would fall, and his employer would be benefited by the difference. But this is no real advantage to his employer; it would indeed operate to increase the rate of his profits, as every fall of wages must do; but in proportion as the labourer contributed less to the fund from which the bounty was paid, and which, let it be remembered, must be raised, his employer must contribute more; in other words, he would contribute as much to the tax by his expenditure as he would receive in the effects of the bounty and the higher rate of profits together. He obtains a higher rate of profits to requite him for his payment, not only of his own quota of the tax, but of his labourer's also; the remuneration which he receives for his labourer's quota appears in diminished wages, or, which is the same thing, in increased profits; the remuneration for his own appears in the diminution in the price of the corn which he consumes, arising from the bounty.

Here it will be proper to remark the different effects produced on profits from an alteration in the real labour, or natural value of corn, and an alteration in the relative value of corn, from taxation and from bounties. If corn is lowered in price by an alteration in its labour price, not only will the rate of the profits of stock be altered, but the condition of the capitalist will be improved. With greater profits, he will have no more to pay

[1] See p. 99.

for the objects on which those profits are expended; which does not happen, as we have just seen, when the fall is occasioned artificially by a bounty. In the real fall in the value of corn, arising from less labour being required to produce one of the most important objects of man's consumption, labour is rendered more productive. With the same capital the same labour is employed, and an increase of productions is the result; not only then will the rate of profits be increased, but the condition of him who obtains them will be improved; not only will each capitalist have a greater money revenue, if he employs the same money capital, but also when that money is expended it will procure him a greater sum of commodities; his enjoyments will be augmented. In the case of the bounty, to balance the advantage which he derives from the fall of one commodity, he has the disadvantage of paying a price more than proportionally high for another; he receives an increased rate of profits in order to enable him to pay this higher price; so that his real situation, though not deteriorated, is in no way improved: though he gets a higher rate of profits, he has no greater command of the produce of the land and labour of the country. When the fall in the value of corn is brought about by natural causes, it is not counteracted by the rise of other commodities; on the contrary, they fall from the raw material falling from which they are made: but when the fall in corn is occasioned by artificial means, it is always counteracted by a real rise in the value of some other commodity, so that if corn be bought cheaper, other commodities are bought dearer.

This, then, is a further proof that no particular disadvantage arises from taxes on necessaries, on account of their raising wages and lowering the rate of profits. Profits are indeed lowered, but only to the amount of the labourer's portion of the tax, which must at all events be paid either by his employer or by the consumer of the produce of the labourer's work. Whether you deduct £50 per annum from the employer's revenue, or add £50 to the prices of the commodities which he consumes, can be of no other consequence to him or to the community than as it may equally affect all other classes. If it be added to the prices of the commodity, a miser may avoid the tax by not consuming; if it be indirectly deducted from every man's revenue, he cannot avoid paying his fair proportion of the public burthens.

A bounty on the production of corn, then, would produce no real effect on the annual produce of the land and labour of the country, although it would make corn relatively cheap and manufactures relatively dear. But suppose now that a contrary measure should be adopted—that a tax should be raised on corn for the purpose of affording a fund for a bounty on the production of commodities.

In such case, it is evident that corn would be dear and commodities cheap; labour would continue at the same price if the labourer were as much benefited by the cheapness of commodities as he was injured by the dearness of corn; but if he were not, wages would rise and profits would fall, while money rent would continue the same as before; profits would fall, because, as we have just explained, that would be the mode in which the labourer's share of the tax would be paid by the employers of labour. By the increase of wages the labourer would be compensated for the tax which he would pay in the increased price of corn; by not expending any part of his wages on the manufactured commodities he would receive no part of the bounty; the bounty would be all received by the employers, and the tax would be partly paid by the employed; a remuneration would be made to the labourers, in the shape of wages, for this increased burden laid upon them, and thus the rate of profits would be reduced. In this case, too, there would be a complicated measure producing no national result whatever.

In considering this question we have purposely left out of our consideration the effect of such a measure on foreign trade; we have rather been supposing the case of an insulated country having no commercial connection with other countries. We have seen that, as the demand of the country for corn and commodities would be the same, whatever direction the bounty might take, there would be no temptation to remove capital from one employment to another; but this would no longer be the case if there were foreign commerce, and that commerce were free. By altering the relative value of commodities and corn, by producing so powerful an effect on their natural prices, we should be applying a strong stimulus to the exportation of those commodities whose natural prices were lowered, and an equal stimulus to the importation of those commodities whose natural prices were raised, and thus such a financial measure might entirely alter the natural distribution of employments, to the advantage indeed of the foreign countries, but ruinously to that in which so absurd a policy was adopted.

CHAPTER XXIV

DOCTRINE OF ADAM SMITH CONCERNING THE RENT OF LAND

" Such parts only of the produce of land," says Adam Smith, " can commonly be brought to market of which the ordinary price is sufficient to replace the stock which must be employed in bringing them thither, together with its ordinary profits. If the ordinary price is more than this, the surplus part of it will naturally go to the rent of land. *If it is not more, though the commodity can be brought to market, it can afford no rent to the landlord.* Whether the price is, or is not more, depends upon the demand."

This passage would naturally lead the reader to conclude that its author could not have mistaken the nature of rent, and that he must have seen that the quality of land which the exigencies of society might require to be taken into cultivation would depend on " *the ordinary price of its produce*" *whether it were* " *sufficient to replace the stock which must be employed in culti-vating it, together with its ordinary profits.*"

But he had adopted the notion that " there were some parts of the produce of land for which the demand must always be such as to afford a greater price than what is sufficient to bring them to market; " and he considered food as one of those parts.

He says that " land, in almost any situation, produces a greater quantity of food than what is sufficient to maintain all the labour necessary for bringing it to market, in the most liberal way in which that labour is ever maintained. The surplus, too, is always more than sufficient to replace the stock which employed that labour, together with its profits. Something, therefore, always remains for a rent to the landlord."

But what proof does he give of this?—no other than the assertion that " the most desert moors in Norway and Scotland produce some sort of pasture for cattle, of which the milk and the increase are always more than sufficient, not only to maintain all the labour necessary for tending them, and to pay the ordinary profit to the farmer, or owner of the herd or flock, but to afford some small rent to the landlord." Now, of this I may be permitted to entertain a doubt; I believe that as yet in every country, from the rudest to the most refined, there is land of such a quality that it cannot yield a produce more than sufficiently valuable to replace the stock employed upon it, together with the profits ordinary and usual in that country. In America we all know that this is the case, and yet no one maintains that the principles which regulate rent are different in that country and in Europe. But if it were true that England had so far advanced in cultivation that at this time there were no lands remaining which did not afford a rent, it would be equally true that there formerly must have been such lands; and that whether there be or not is of no importance to this question, for it is the same thing if there be any capital employed in Great Britain on land which yields only the return of stock with its ordinary profits, whether it be employed on old or on new land. If a farmer agrees for land on a lease of seven or fourteen years, he may propose to employ on it a capital of £10,000, knowing that at the existing price of grain and raw produce he can replace that part of his stock which he is obliged to expend, pay his rent, and obtain the general rate of profit. He will not employ £11,000, unless the last £1000 can be employed so productively as to afford him the usual profits of stock. In his calculation, whether he shall employ it or not, he considers only whether the price of raw produce is sufficient to replace his expenses and profits, for he knows that he shall have no additional rent to pay. Even at the expiration of his lease his rent will not be raised; for if his landlord should require rent, because this additional £1000 was employed, he would withdraw it; since, by employing it, he gets, by the supposition, only the ordinary and usual profits which he may obtain by any other employment of stock; and, therefore, he cannot afford to pay rent for it, unless the price of raw produce should further rise, or, which is the same thing, unless the usual and general rate of profits should fall.

If the comprehensive mind of Adam Smith had been directed to this fact, he would not have maintained that rent forms one of the component parts of the price of raw produce; for price is everywhere regulated by the return obtained by this last portion of capital, for which no rent whatever is paid. If he had adverted to this principle, he would have made no distinction between the law which regulates the rent of mines and the rent of land.

"Whether a coal mine, for example," he says, "can afford any rent depends partly upon its fertility and partly upon its situation. A mine of any kind may be said to be either fertile or barren according as the quantity of mineral which can be brought from it by a certain quantity of labour is greater or less than what can be brought by an equal quantity from the greater part of other mines of the same kind. Some coal mines, advantageously situated, cannot be wrought on account of their barrenness. The produce does not pay the expense. They can afford neither profit nor rent. There are some of which the produce is barely sufficient to pay the labour and replace, together with its ordinary profits, the stock employed in working them. They afford some profit to the undertaker of the work, but no rent to the landlord. They can be wrought advantageously by nobody but the landlord, who being himself the undertaker of the work, gets the ordinary profit of the capital which he employs in it. Many coal mines in Scotland are wrought in this manner, and can be wrought in no other. The landlord will allow nobody else to work them without paying some rent, and nobody can afford to pay any.

" Other coal mines in the same country, sufficiently fertile, cannot be wrought on account of their situation. A quantity of mineral sufficient to defray the expense of working could be brought from the mine by the ordinary, or even less than the ordinary, quantity of labour; but in an inland country, thinly inhabited, and without either good roads or water-carriage, this quantity could not be sold." The whole principle of rent is here admirably and perspicuously explained, but every word is as applicable to land as it is to mines; yet he affirms that " it is otherwise in estates above ground. The proportion, both of their produce and of their rent, is in proportion to their absolute, and not to their relative, fertility." But, suppose that there were no land which did not afford a rent; then the amount of rent on the worst land would be in proportion to the excess of the value of the produce above the expenditure of capital and the ordinary profits of stock: the same principle would govern the rent of land of a somewhat better quality, or more favourably situated, and, therefore, the rent of this land would exceed the rent of that inferior to it by the superior advantages which it possessed; the same might be said of that of the third quality, and so on to the very best. Is it not, then, as certain that it is the relative fertility of the land which determines the portion of the produce which shall be paid for the rent of land as it is that the relative fertility of mines determines the portion of their produce which shall be paid for the rent of mines?

After Adam Smith has declared that there are some mines which can only be worked by the owners, as they will afford only sufficient to defray the expense of working, together with the ordinary profits of the capital employed, we should expect that he would admit that it was these particular mines which regulated the price of the produce from all mines. If the old mines are insufficient to supply the quantity of coal required, the price of coal will rise, and will continue rising till the owner of a new and inferior mine finds that he can obtain the usual profits of stock by working his mine. If his mine be tolerably fertile, the rise will not be great before it becomes his interest so to employ his capital; but if it be not tolerably fertile, it is evident that the price must continue to rise till it will afford him the means of paying his expenses, and obtaining the ordinary profits of stock. It appears, then, that it is always the least fertile mine which regulates the price of coal. Adam Smith, however, is of a different opinion: he observes that " the most fertile coal mine, too, regulates the price of coals at all the other mines in its neighbourhood. Both the proprietor and the undertaker of the work find, the one that he can get a greater rent, the other that he can get a greater profit, by somewhat underselling all their neighbours. Their neighbours are soon obliged to sell at the same price, though they cannot so well afford it, and though it always diminishes, and sometimes takes away altogether, both their rent and their profit. Some works are abandoned altogether; others can afford no rent, and can be wrought only by the proprietor." If the demand for coal should be diminished, or if by new processes the quantity should be increased, the price would fall, and some mines would be abandoned; but in every case, the price must be sufficient to pay the expenses and profit of that mine which is worked without being charged with rent. It is, therefore, the least fertile mine which regulates price. Indeed, it is so stated in another place by Adam Smith himself, for he says, " The lowest price at which coals can be sold for any considerable time is like that of all other commodities, the price which is barely sufficient to replace, together with its ordinary profits, the stock which must be employed in bringing them to market. At a coal mine for which the landlord can get no rent, but which he must either work himself, or let it alone all altogether, the price of coals must generally be nearly about this price."

But the same circumstance, namely, the abundance and consequent cheapness of coals, from whatever cause it may arise, which would make it necessary to abandon those mines on which there was no rent, or a very moderate one, would, if there were the same abundance and consequent cheapness of raw produce, render it necessary to abandon the cultivation of those lands for which either no rent was paid or a very moderate one. If, for example, potatoes should become the general and common food of the people, as rice is in some countries, one-fourth or one-half of the land now in cultivation would probably be immediately abandoned; for if, as Adam Smith says, " an acre of potatoes will produce six thousand weight of solid nourishment, three times the quantity produced by the acre of wheat," there could not be for a considerable time such a multiplication of people as to consume the quantity that might be raised on the land before employed for the cultivation of wheat; much land would consequently be abandoned, and rent would fall; and it would not be till the population had been doubled or trebled that the same quantity of land could be in cultivation and the rent paid for it as high as before.

Neither would any greater proportion of the gross produce be paid to the landlord whether it consisted of potatoes, which would feed three hundred people, or of wheat, which would feed only one hundred; because, though the expenses of production would be very much diminished if the labourer's wages were chiefly regulated by the price of potatoes, and not by the price of wheat, and though, therefore, the proportion of the whole gross produce, after paying the labourers, would be greatly increased, yet no part of that additional proportion would go to rent, but the whole invariably to profits—profits being at all times raised as wages fall, and lowered as wages rise. Whether wheat or potatoes were cultivated, rent would be governed by the same principle—it would be always equal to the difference between the quantities of produce obtained with equal capitals, either on the same land or on land of different qualities; and, therefore, while lands of the same quality were cultivated, and there was no alteration in their relative fertility or advantages, rent would always bear the same proportion to the gross produce.

Adam Smith, however, maintains that the proportion which falls to the landlord would be increased by a diminished cost of production, and, therefore, that he would receive a larger share as well as a larger quantity from an abundant than from a scanty produce. " A rice field," he says, " produces a much greater quantity of food than the most fertile corn field. Two

crops in the year, from thirty to sixty bushels each, are said
to be the ordinary produce of an acre. Though its cultivation,
therefore, requires more labour, a much greater surplus remains
after maintaining all that labour. In those rice countries,
therefore, where rice is the common and favourite vegetable
food of the people, and where the cultivators are chiefly maintained with it, *a greater share of this greater surplus should belong
to the landlord than in corn countries.*"

Mr. Buchanan also remarks that " it is quite clear that if any
other produce, which the land yielded more abundantly than
corn, were to become the common food of the people, the rent
of the landlord would be improved in proportion to its greater
abundance."

If potatoes were to become the common food of the people,
there would be a long interval during which the landlords would
suffer an enormous deduction of rent. They would not probably
receive nearly so much of the sustenance of man as they now
receive, while that sustenance would fall to a third of its present
value. But all manufactured commodities, on which a part
of the landlord's rent is expended, would suffer no other fall
than that which proceeded from the fall in the raw material
of which they were made, and which would arise only from the
greater fertility of the land which might then be devoted to its
production.

When, from the progress of population, land of the same
quality as before should be taken into cultivation, the landlord
would have not only the same proportion of the produce as
before, but that proportion would also be of the same value as
before. Rent, then, would be the same as before; profits,
however, would be much higher, because the price of food, and
consequently wages, would be much lower. High profits are
favourable to the accumulation of capital. The demand for
labour would further increase, and landlords would be permanently benefited by the increased demand for land.

Indeed, the very same lands might be cultivated much higher
when such an abundance of food could be produced from them,
and, consequently, they would, in the progress of society, admit
of much higher rents, and would sustain a much greater population than before. This could not fail to be highly beneficial to
landlords, and is consistent with the principle which this inquiry,
I think, will not fail to establish—that all extraordinary profits
are in their nature but of limited duration, as the whole surplus
produce of the soil, after deducting from it only such moderate

profits as are sufficient to encourage accumulation, must finally rest with the landlord.

With so low a price of labour as such an abundant produce would cause, not only would the lands already in cultivation yield a much greater quantity of produce, but they would admit of a great additional capital being employed on them, and a greater value to be drawn from them, and, at the same time, lands of a very inferior quality could be cultivated with high profits, to the great advantage of landlords, as well as to the whole class of consumers. The machine which produced the most important article of consumption would be improved, and would be well paid for according as its services were demanded. All the advantages would, in the first instance, be enjoyed by labourers, capitalists, and consumers; but, with the progress of population, they would be gradually transferred to the proprietors of the soil.

Independently of these improvements, in which the community have an immediate and the landlords a remote interest, the interest of the landlord is always opposed to that of the consumer and manufacturer. Corn can be permanently at an advanced price only because additional labour is necessary to produce it; because its cost of production is increased. The same cause invariably raises rent, it is therefore for the interest of the landlord that the cost attending the production of corn should be increased. This, however, is not the interest of the consumer; to him it is desirable that corn should be low relatively to money and commodities, for it is always with commodities or money that corn is purchased. Neither is it the interest of the manufacturer that corn should be at a high price, for the high price of corn will occasion high wages, but will not raise the price of his commodity. Not only, then, must more of his commodity, or, which comes to the same thing, the value of more of his commodity, be given in exchange for the corn which he himself consumes, but more must be given, or the value of more, for wages to his workmen, for which he will receive no remuneration. All classes, therefore, except the landlords, will be injured by the increase in the price of corn. The dealings between the landlord and the public are not like dealings in trade, whereby both the seller and buyer may equally be said to gain, but the loss is wholly on one side, and the gain wholly on the other; and if corn could by importation be procured cheaper, the loss in consequence of not importing is far greater on one side than the gain is on the other.

Adam Smith never makes any distinction between a low value of money and a high value of corn, and therefore infers that the interest of the landlord is not opposed to that of the rest of the community. In the first case, money is low relatively to all commodities; in the other, corn is high relatively to all. In the first, corn and commodities continue at the same relative values; in the second, corn is higher relatively to commodities as well as money.

The following observation of Adam Smith is applicable to a low value of money, but it is totally inapplicable to a high value of corn. " If importation (of corn) was at all times free, our farmers and country gentlemen would probably, one year with another, get less money for their corn than they do at present when importation is at most times in effect prohibited; but the money which they got would be of more value, *would buy more goods of all other kinds*, and would employ more labour. Their real wealth, their real revenue, therefore, would be the same as at present, though it might be expressed by a smaller quantity of silver; and they would neither be disabled nor discouraged from cultivating corn as much as they do at present. On the contrary, as the rise in the real value of silver, in consequence of lowering the money price of corn, lowers somewhat the money price of all other commodities, it gives the industry of the country where it takes place some advantage in all foreign markets, and thereby tends to encourage and increase that industry. But the extent of the home market for corn must be in proportion to the general industry of the country where it grows, or to the number of those who produce something else to give in exchange for corn. But in every country the home market, as it is the nearest and most convenient, so is it likewise the greatest and most important market for corn. That rise in the real value of silver, therefore, which is the effect of lowering the average money price of corn, tends to enlarge the greatest and most important market for corn, and thereby to encourage instead of discouraging its growth."

A high or low money price of corn, arising from the abundance and cheapness of gold and silver, is of no importance to the landlord, as every sort of produce would be equally affected just as Adam Smith describes; but a relatively high price of corn is at all times greatly beneficial to the landlord; for, first, it gives him a greater quantity of corn for rent; and, secondly, for every equal measure of corn he will have a command, not only over a greater quantity of money, but over a greater quantity of every commodity which money can purchase,

CHAPTER XXV

ON COLONIAL TRADE

ADAM SMITH, in his observations on colonial trade, has shown most satisfactorily the advantages of a free trade, and the injustice suffered by colonies in being prevented by their mother countries from selling their produce at the dearest market and buying their manufactures and stores at the cheapest. He has shown that, by permitting every country freely to exchange the produce of its industry when and where it pleases, the best distribution of the labour of the world will be effected, and the greatest abundance of the necessaries and enjoyments of human life will be secured.

He has attempted also to show that this freedom of commerce, which undoubtedly promotes the interest of the whole, promotes also that of each particular country; and that the narrow policy adopted in the countries of Europe respecting their colonies is not less injurious to the mother countries themselves than to the colonies whose interests are sacrificed.

" The monopoly of the colony trade," he says, " like all the other mean and malignant expedients of the mercantile system, depresses the industry of all other countries, but chiefly that of the colonies, without in the least increasing, but, on the contrary, diminishing that of the country in whose favour it is established."

This part of his subject, however, is not treated in so clear and convincing a manner as that in which he shows the injustice of this system towards the colony.

It may, I think, be doubted whether a mother country may not sometimes be benefited by the restraints to which she subjects her colonial possessions. Who can doubt, for example, that if England were the colony of France, the latter country would be benefited by a heavy bounty paid by England on the exportation of corn, cloth, or any other commodities? In examining the question of bounties, on the supposition of corn being at £4 per quarter in this country, we saw that with a bounty of 10s. per quarter on exportation in England, corn would have been reduced to £3 10s. in France. Now, if corn had previously been at £3 15s. per quarter in France, the French

227

consumers would have been benefited by 5s. per quarter on all imported corn; if the natural price of corn in France were before £4, they would have gained the whole bounty of 10s. per quarter. France would thus be benefited by the loss sustained by England: she would not gain a part only of what England lost, but the whole.

It may, however, be said that a bounty on exportation is a measure of internal policy, and could not easily be imposed by the mother country.

If it would suit the interests of Jamaica and Holland to make an exchange of the commodities which they respectively produce, without the intervention of England, it is quite certain that by their being prevented from so doing the interests of Holland and Jamaica would suffer; but if Jamaica is obliged to send her goods to England, and there exchange them for Dutch goods, an English capital, or English agency, will be employed in a trade in which it would not otherwise be engaged. It is allured thither by a bounty, not paid by England, but by Holland and Jamaica.

That the loss sustained through a disadvantageous distribution of labour in two countries may be beneficial to one of them, while the other is made to suffer more than the loss actually belonging to such a distribution, has been stated by Adam Smith himself; which, if true, will at once prove that a measure which may be greatly hurtful to a colony may be partially beneficial to the mother country.

Speaking of treaties of commerce, he says, " When a nation binds itself by treaty, either to permit the entry of certain goods from one foreign country which it prohibits from all others, or to exempt the goods of one country from duties to which it subjects those of all others, the country, or at least the merchants and manufacturers of the country, whose commerce is so favoured, must necessarily derive great advantage from the treaty. Those merchants and manufacturers enjoy a sort of monopoly in the country which is so indulgent to them. That country becomes a market both more extensive and more advantageous for their goods; more extensive, because the goods of other nations, being either excluded or subjected to heavier duties, it takes off a greater quantity of them; more advantageous, because the merchants of the favoured country, enjoying a sort of monopoly there, will often sell their goods for a better price than if exposed to the free competition of all other nations."

Let the two nations between which the commercial treaty is made be the mother country and her colony, and Adam Smith, it is evident, admits that a mother country may be benefited by oppressing her colony. It may, however, be again remarked, that unless the monopoly of the foreign market be in the hands of an exclusive company, no more will be paid for commodities by foreign purchasers than by home purchasers; the price which they will both pay will not differ greatly from their natural price in the country where they are produced. England, for example, will, under ordinary circumstances, always be able to buy French goods at the natural price of those goods in France, and France would have an equal privilege of buying English goods at their natural price in England. But at these prices goods would be bought without a treaty. Of what advantage or disadvantage, then, is the treaty to either party?

The disadvantage of the treaty to the importing country would be this: it would bind her to purchase a commodity, from England, for example, at the natural price of that commodity in England, when she might perhaps have bought it at the much lower natural price of some other country. It occasions then a disadvantageous distribution of the general capital, which falls chiefly on the country bound by its treaty to buy in the least productive market; but it gives no advantage to the seller on account of any supposed monopoly, for he is prevented by the competition of his own countrymen from selling his goods above their natural price; at which he would sell them, whether he exported them to France, Spain, or the West Indies, or sold them for home consumption.

In what, then, does the advantage of the stipulation in the treaty consist? It consists in this: these particular goods could not have been made in England for exportation, but for the privilege which she alone had of serving this particular market; for the competition of that country, where the natural price was lower, would have deprived her of all chance of selling those commodities. This, however, would have been of little importance if England were quite secure that she could sell to the same amount any other goods which she might fabricate, either in the French market or with equal advantage in any other. The object which England has in view is, for example, to buy a quantity of French wines of the value of £5000—she desires, then, to sell goods somewhere by which she may get £5000 for this purpose. If France gives her a monopoly of the cloth market she will readily export cloth for this purpose; but if the trade is free, the competition of other countries may prevent the natural price of cloth in England from being sufficiently low to enable her to get £5000 by the sale of cloth, and to obtain the usual profits by such an employment of her stock. The industry of England must be employed, then, on some other commodity; but there may be none of her productions which, at the existing value of money, she can afford to sell at the natural price of other countries. What is the consequence? The wine drinkers of England are still willing to give £5000 for their wine, and consequently £5000 in money is exported to France for that purpose. By this exportation of money, its value is raised in England and lowered in other countries; and with it the *natural price* of all commodities produced by British industry is also lowered. The advance in the value of money is the same thing as the decline in the price of commodities. To obtain £5000, British commodities may now be exported; for at their reduced natural price they may now enter into competition with the goods of other countries. More goods are sold, however, at the low prices to obtain the £5000 required, which, when obtained, will not procure the same quantity of wine; because, whilst the diminution of money in England has lowered the natural price of goods there, the increase of money in France has raised the natural price of goods and wine in France. Less wine, then, will be imported into England, in exchange for its commodities, when the trade is perfectly free than when she is peculiarly favoured by commercial treaties. The *rate* of profits, however, will not have varied; money will have altered in relative value in the two countries, and the advantage gained by France will be the obtaining a greater quantity of English, in exchange for a given quantity of French, goods, while the loss sustained by England will consist in obtaining a smaller quantity of French goods in exchange for a given quantity of those of England.

Foreign trade, then, whether fettered, encouraged, or free, will always continue, whatever may be the comparative difficulty of production in different countries; but it can only be regulated by altering the natural price, not the natural value, at which commodities can be produced in those countries, and that is effected by altering the distribution of the precious metals. This explanation confirms the opinion which I have elsewhere given, that there is not a tax, a bounty, or a prohibition on the importation or exportation of commodities which does not occasion a different distribution of the precious metals, and which does not, therefore, everywhere alter both the natural and the market price of commodities.

It is evident, then, that the trade with a colony may be so regulated that it shall at the same time be less beneficial to the colony, and more beneficial to the mother country, than a perfectly free trade. As it is disadvantageous to a single consumer to be restricted in his dealings to one particular shop, so is it disadvantageous for a nation of consumers to be obliged to purchase of one particular country. If the shop or the country afforded the goods required the cheapest, they would be secure of selling them without any such exclusive privilege; and if they did not sell cheaper, the general interest would require that they should not be encouraged to continue a trade which they could not carry on at an equal advantage with others. The shop, or the selling country, might lose by the change of employments, but the general benefit is never so fully secured as by the most productive distribution of the general capital; that is to say, by a universally free trade.

An increase in the cost of production of a commodity, if it be an article of the first necessity, will not necessarily diminish its consumption; for although the general power of the purchasers to consume is diminished by the rise of any one commodity, yet they may relinquish the consumption of some other commodity whose cost of production has not risen. In that case, the quantity supplied, and the quantity demanded, will be the same as before; the cost of production only will have increased, and yet the price will rise, and must rise, to place the profits of the producer of the enhanced commodity on a level with the profits derived from other trades.

M. Say acknowledges that the cost of production is the foundation of price, and yet in various parts of his book he maintains that price is regulated by the proportion which demand bears to supply. The real and ultimate regulator of the relative value of any two commodities is the cost of their production, and not the respective quantities which may be produced, nor the competition amongst the purchasers.

According to Adam Smith, the colony trade, by being one in which British capital only can be employed, has raised the rate of profits of all other trades; and as, in his opinion, high profits, as well as high wages, raise the prices of commodities, the monopoly of the colony trade has been, he thinks, injurious to the mother country; as it has diminished her power of selling manufactured commodities as cheap as other countries. He says that " in consequence of the monopoly, the increase of the colony trade has not so much occasioned an addition to the trade which Great Britain had before as a total change in its direction. Secondly, this monopoly has necessarily contributed to keep up the rate of profit in all the different branches of British trade higher than it naturally would have been had all nations been allowed a free trade to the British colonies."

" But whatever raises in any country the ordinary rate of profit higher than it otherwise would be, necessarily subjects that country both to an absolute and to a relative disadvantage in every branch of trade of which she has not the monopoly. It subjects her to an absolute disadvantage, because in such branches of trade her merchants cannot get this greater profit without selling dearer than they otherwise would do both the goods of foreign countries which they import into their own and the goods of their own country which they export to foreign countries. Their own country must both buy dearer and sell dearer; must both buy less and sell less; must both enjoy less and produce less than she otherwise would do."

" Our merchants frequently complain of the high wages of British labour as the cause of their manufactures being undersold in foreign markets; but they are silent about the high profits of stock. They complain of the extravagant gain of other people, but they say nothing of their own. The high profits of British stock, however, may contribute towards raising the price of British manufacture in many cases as much, and in some perhaps more, than the high wages of British labour."

I allow that the monopoly of the colony trade will change, and often prejudicially, the direction of capital; but from what I have already said on the subject of profits, it will be seen that any change from one foreign trade to another, or from home to foreign trade, cannot, in my opinion, affect the rate of profits. The injury suffered will be what I have just described; there will be a worse distribution of the general capital and industry, and, therefore, less will be produced. The natural price of commodities will be raised, and therefore, though the consumer will be able to purchase to the same money value, he will obtain a less quantity of commodities. It will be seen, too, that if it even had the effect of raising profits, it would not occasion the least alteration in prices; prices being regulated neither by wages nor profits.

And does not Adam Smith agree in this opinion, when he says

that " the prices of commodities, or the value of gold and silver as compared with commodities, depends upon the proportion between the *quantity of labour* which is necessary in order to bring a certain quantity of gold and silver to market, and that which is necessary to bring thither a certain quantity of any other sort of goods? " That quantity will not be affected, whether profits be high or low, or wages low or high. How then can prices be raised by high profits?

CHAPTER XXVI

ON GROSS AND NET REVENUE

ADAM SMITH constantly magnifies the advantages which a country derives from a large gross, rather than a large net income. " In proportion as a greater share of the capital of a country is employed in agriculture," he says, " the greater will be the quantity of productive labour which it puts into motion within the country; as will likewise be the value which its employment adds to the annual produce of the land and labour of the society. After agriculture, the capital employed in manufactures puts into motion the greatest quantity of productive labour, and adds the greatest value to the annual produce. That which is employed in the trade of exportation has the least effect of any of the three." [1]

Granting, for a moment, that this were true, what would be the advantage resulting to a country from the employment of a great quantity of productive labour, if, whether it employed that quantity or a smaller, its net rent and profits together would be the same. The whole produce of the land and labour of every country is divided into three portions: of these, one portion is devoted to wages, another to profits, and the other to rent. It is from the two last portions only that any deductions can be made for taxes or for savings; the former, if moderate, constituting always the necessary expenses of production. [2] To an individual with a capital of £20,000, whose profits were £2000 per annum, it would be a matter quite indifferent whether

---

[1] M. Say is of the same opinion with Adam Smith: " The most productive employment of capital, for the country in general, after that on the land, is that of manufactures and of home trade; because it puts in activity an industry of which the profits are gained in the country, while those capitals which are employed in foreign commerce make the industry and lands of all countries to be productive, without distinction.

" The employment of capital the least favourable to a nation is that of carrying the produce of one foreign country to another."—Say, vol. ii. p. 120.

[2] Perhaps this is expressed too strongly, as more is generally allotted to the labourer under the name of wages than the absolutely necessary expenses of production. In that case a part of the net produce of the country is received by the labourer, and may be saved or expended by him; or it may enable him to contribute to the defence of the country.

his capital would employ a hundred or a thousand men, whether the commodity produced sold for £10,000 or for £20,000, provided, in all cases, his profits were not diminished below £2000. Is not the real interest of the nation similar? Provided its net real income, its rent and profits be the same, it is of no importance whether the nation consists of ten or of twelve millions of inhabitants. Its power of supporting fleets and armies, and all species of unproductive labour, must be in proportion to its net, and not in proportion to its gross, income. If five millions of men could produce as much food and clothing as was necessary for ten millions, food and clothing for five millions would be the net revenue. Would it be of any advantage to the country that, to produce this same net revenue, seven millions of men should be required, that is to say, that seven millions should be employed to produce food and clothing sufficient for twelve millions? The food and clothing of five millions would be still the net revenue. The employing a greater number of men would enable us neither to add a man to our army and navy, nor to contribute one guinea more in taxes.

It is not on the grounds of any supposed advantage accruing from a large population, or of the happiness that may be enjoyed by a greater number of human beings, that Adam Smith supports the preference of that employment of capital which gives motion to the greatest quantity of industry, but expressly on the ground of its increasing the power of the country,[1] for he says that " the riches and, so far as power depends upon riches, the power of every country must always be in proportion to the value of its annual produce, the fund from which all taxes must ultimately be paid." It must, however, be obvious that the power of paying taxes is in proportion to the net, and not in proportion to the gross, revenue.

In the distribution of employments amongst all countries, the capital of poorer nations will be naturally employed in those pursuits wherein a great quantity of labour is supported at home, because in such countries the food and necessaries for an increasing population can be most easily procured. In rich countries, on the contrary, where food is dear, capital will naturally flow, when trade is free, into those occupations wherein

---

[1] M. Say has totally misunderstood me in supposing that I have considered as nothing the happiness of so many human beings. I think the text sufficiently shows that I was confining my remarks to the particular grounds on which Adam Smith had rested it.

the least quantity of labour is required to be maintained at home: such as the carrying trade, the distant foreign trade, and trades where expensive machinery is required; to trades where profits are in proportion to the capital, and not in proportion to the quantity of labour employed.[1]

Although I admit that, from the nature of rent, a given capital employed in agriculture, on any but the land last cultivated, puts in motion a greater quantity of labour than an equal capital employed in manufactures and trade, yet I cannot admit that there is any difference in the quantity of labour employed by a capital engaged in the home trade and an equal capital engaged in the foreign trade.

" The capital which sends Scotch manufactures to London, and brings back English corn and manufactures to Edinburgh," says Adam Smith, " necessarily replaces, by every such operation, two British capitals which had both been employed in the agriculture or manufactures of Great Britain.

" The capital employed in purchasing foreign goods for home consumption, when this purchase is made with the produce of domestic industry, replaces, too, by every such operation, two distinct capitals; but one of them only is employed in supporting domestic industry. The capital which sends British goods to Portugal, and brings back Portuguese goods to Great Britain, replaces, by every such operation, only one British capital, the other is a Portuguese one. Though the returns, therefore, of the foreign trade of consumption should be as quick as the home trade, the capital employed in it will give but one half the encouragement to the industry or productive labour of the country."

This argument appears to me to be fallacious; for though two capitals, one Portuguese and one English, be employed, as Dr. Smith supposes, still a capital will be employed in the foreign trade double of what would be employed in the home trade. Suppose that Scotland employs a capital of a thousand pounds in making linen, which she exchanges for the produce of a similar capital employed in making silks in England, two

---

[1] " It is fortunate that the natural course of things draws capital, not to those employments where the greatest profits are made, but to those where the operation is most profitable to the community."—Vol. ii. p. 122. M. Say has not told us what those employments are which, while they are the most profitable to the individual, are not the most profitable to the state. If countries with limited capitals, but with abundance of fertile land, do not early engage in foreign trade, the reason is, because it is less profitable to individuals, and therefore also less profitable to the state.

thousand pounds and a proportional quantity of labour will be employed by the two countries. Suppose now that England discovers that she can import more linen from Germany for the silks which she before exported to Scotland, and that Scotland discovers that she can obtain more silks from France in return for her linen than she before obtained from England, will not England and Scotland immediately cease trading with each other, and will not the home trade of consumption be changed for a foreign trade of consumption? But although two additional capitals will enter into this trade, the capital of Germany and that of France, will not the same amount of Scotch and of English capital continue to be employed, and will it not give motion to the same quantity of industry as when it was engaged in the home trade?

CHAPTER XXVII

ON CURRENCY AND BANKS

So much has already been written on currency that of those who give their attention to such subjects none but the prejudiced are ignorant of its true principles. I shall, therefore, take only a brief survey of some of the general laws which regulate its quantity and value.

Gold and silver, like all other commodities, are valuable only in proportion to the quantity of labour necessary to produce them and bring them to market. Gold is about fifteen times dearer than silver, not because there is a greater demand for it, nor because the supply of silver is fifteen times greater than that of gold, but solely because fifteen times the quantity of labour is necessary to procure a given quantity of it.

The quantity of money that can be employed in a country must depend on its value: if gold alone were employed for the circulation of commodities, a quantity would be required one fifteenth only of what would be necessary if silver were made use of for the same purpose.

A circulation can never be so abundant as to overflow; for by diminishing its value in the same proportion you will increase its quantity, and by increasing its value, diminish its quantity.

While the state coins money, and charges no seignorage, money will be of the same value as any other piece of the same metal of equal weight and fineness; but if the state charges a seignorage for coinage, the coined piece of money will generally exceed the value of the uncoined piece of metal by the whole seignorage charged, because it will require a greater quantity of labour, or, which is the same thing, the value of the produce of a greater quantity of labour, to procure it.

While the state alone coins, there can be no limit to this charge of seignorage; for by limiting the quantity of coin, it can be raised to any conceivable value.

It is on this principle that paper money circulates: the whole charge for paper money may be considered as seignorage. Though it has no intrinsic value, yet, by limiting its quantity, its value in exchange is as great as an equal denomination of coin, or of bullion in that coin. On the same principle, too, namely, by a limitation of its quantity, a debased coin would circulate at the value it should bear if it were of the legal weight and fineness, and not at the value of the quantity of metal which it actually contained. In the history of the British coinage we find, accordingly, that the currency was never depreciated in the same proportion that it was debased; the reason of which was, that it never was increased in quantity in proportion to its diminished intrinsic value.[1]

There is no point more important in issuing paper money than to be fully impressed with the effects which follow from the principle of limitation of quantity. It will scarcely be believed fifty years hence that bank directors and ministers gravely contended in our times, both in Parliament and before committees of Parliament, that the issues of notes by the Bank of England, unchecked by any power in the holders of such notes to demand in exchange either specie or bullion, had not, nor could have, any effect on the prices of commodities, bullion, or foreign exchanges.

After the establishment of banks, the state has not the sole power of coining or issuing money. The currency may as effectually be increased by paper as by coin; so that if a state were to debase its money, and limit its quantity, it could not support its value, because the banks would have an equal power of adding to the whole quantity of circulation.

On these principles, it will be seen that it is not necessary that paper money should be payable in specie to secure its value; it is only necessary that its quantity should be regulated according to the value of the metal which is declared to be the standard. If the standard were gold of a given weight and fineness, paper might be increased with every fall in the value of gold, or, which is the same thing in its effects, with every rise in the price of goods.

"By issuing too great a quantity of paper," says Dr. Smith, "of which the excess was continually returning in order to be exchanged for gold and silver, the Bank of England was, for many years together, obliged to coin gold to the extent of between eight hundred thousand pounds and a million a year, or, at an average, about eight hundred and fifty thousand pounds. For this great coinage, the Bank, in consequence of the worn and degraded state into which the gold coin had fallen

---

[1] Whatever I say of gold coin is equally applicable to silver coin; but it is not necessary to mention both on every occasion.

a few years ago, was frequently obliged to purchase bullion at the high price of four pounds an ounce, which it soon after issued in coin at £3 17s. 10½d. an ounce, losing in this manner between two and a half and three per cent. upon the coinage of so very large a sum. Though the Bank, therefore, paid no seignorage, though the government was properly at the expense of the coinage, this liberality of government did not prevent altogether the expense of the Bank."

On the principle above stated, it appears to me most clear that by not re-issuing the paper thus brought in, the value of the whole currency, of the degraded as well as the new gold coin, would have been raised, when all demands on the Bank would have ceased.

Mr. Buchanan, however, is not of this opinion, for he says "that the great expense to which the Bank was at this time exposed was occasioned, not as Dr. Smith seems to imagine, by an imprudent issue of paper, but by the debased state of the currency and the consequent high price of bullion. The Bank, it will be observed, having no other way of procuring guineas but by sending bullion to the Mint to be coined, was always forced to issue new coined guineas in exchange for its returned notes; and when the currency was generally deficient in weight, and the price of bullion high in proportion, it became profitable to draw these heavy guineas from the bank in exchange for its paper; to convert them into bullion, and to sell them with a profit for Bank paper, to be again returned to the Bank for a new supply of guineas, which were again melted and sold. To this drain of specie the Bank must always be exposed while the currency is deficient in weight, as both an easy and a certain profit then arises from the constant interchange of paper for specie. It may be remarked, however, that to whatever inconvenience and expense the Bank was then exposed by the drain of its specie, it never was imagined necessary to rescind the obligation to pay money for its notes."

Mr. Buchanan evidently thinks that the whole currency must necessarily be brought down to the level of the value of the debased pieces; but surely, by a diminution of the quantity of the currency, the whole that remains can be elevated to the value of the best pieces.

Dr. Smith appears to have forgotten his own principle in his argument on colony currency. Instead of ascribing the depreciation of that paper to its too great abundance, he asks whether, allowing the colony security to be perfectly good, a hundred pounds, payable fifteen years hence, would be equally valuable with a hundred pounds to be paid immediately? I answer yes, if it be not too abundant.

Experience, however, shows that neither a state nor a bank ever have had the unrestricted power of issuing paper money without abusing that power; in all states, therefore, the issue of paper money ought to be under some check and control; and none seems so proper for that purpose as that of subjecting the issuers of paper money to the obligation of paying their notes either in gold coin or bullion.

[" To secure the public [1] against any other variations in the value of currency than those to which the standard itself is subject, and, at the same time, to carry on the circulation with a medium the least expensive, is to attain the most perfect state to which a currency can be brought, and we should possess all these advantages by subjecting the Bank to the delivery of uncoined gold or silver at the Mint standard and price, in exchange for their notes, instead of the delivery of guineas; by which means paper would never fall below the value of bullion without being followed by a reduction of its quantity. To prevent the rise of paper above the value of bullion, the Bank should be also obliged to give their paper in exchange for standard gold at the price of £3 17s. per ounce. Not to give too much trouble to the Bank, the quantity of gold to be demanded in exchange for paper at the Mint price of £3 17s. 10½d., or the quantity to be sold to the Bank at £3 17s., should never be less than twenty ounces. In other words, the Bank should be obliged to purchase any quantity of gold that was offered them, not less than twenty ounces, at £3 17s.[2] per ounce, and to sell any quantity that might be demanded at £3 17s. 10½d. While they have the power of regulating the quantity of their paper there is no possible inconvenience that could result to them from such a regulation.

" The most perfect liberty should be given, at the same time, to export or import every description of bullion. These trans-

---

[1] This, and the following paragraphs, to the close of the bracket, p. 244, is extracted from a pamphlet entitled *Proposals for an Economical and Secure Currency*, published by the author in the year 1816.

[2] The price of £3 17s. here mentioned is of course an arbitrary price. There might be good reason, perhaps, for fixing it either a little above or a little below. In naming £3 17s., I wish only to elucidate the principle. The price ought to be so fixed as to make it the interest of the seller of gold rather to sell it to the Bank than to carry it to the Mint to be coined.

The same remark applies to the specified quantity of twenty ounces. There might be good reason for making it ten or thirty.

actions in bullion would be very few in number, if the Bank regulated their loans and issues of paper by the criterion which I have so often mentioned, namely, the price of standard bullion, without attending to the absolute quantity of paper in circulation.

" The object which I have in view would be in a great measure attained if the Bank were obliged to deliver uncoined bullion, in exchange for their notes, at the Mint price and standard, though they were not under the necessity of purchasing any quantity of bullion offered them at the prices to be fixed, particularly if the Mint were to continue open to the public for the coinage of money; for that regulation is merely suggested to prevent the value of money from varying from the value of bullion more than the trifling difference between the prices at which the Bank should buy and sell, and which would be an approximation to that uniformity in its value which is acknowledged to be so desirable.

" If the Bank capriciously limited the quantity of their paper they would raise its value, and gold might appear to fall below the limits at which I propose the Bank should purchase. Gold, in that case, might be carried to the Mint, and the money returned from thence, being added to the circulation, would have the effect of lowering its value, and making it again conform to the standard; but it would neither be done so safely, so economically, nor so expeditiously as by the means which I have proposed, against which the Bank can have no objection to offer, as it is for their interest to furnish the circulation with paper rather than oblige others to furnish it with coin.

" Under such a system, and with a currency so regulated, the bank would never be liable to any embarrassments whatever, excepting on those extraordinary occasions when a general panic seizes the country, and when every one is desirous of possessing the precious metals as the most convenient mode of realising or concealing his property. Against such panics banks have no security *on any system ;* from their very nature they are subject to them, as at no time can there be in a bank, or in a country, so much specie or bullion as the moneyed individuals of such country have a right to demand. Should every man withdraw his balance from his banker on the same day, many times the quantity of bank notes now in circulation would be insufficient to answer such a demand. A panic of this kind was the cause of the crisis in 1797; and not, as has been supposed, the large advances which the Bank had then made to government. Neither the Bank nor government were at that time to blame; it was the contagion of the unfounded fears of the timid part of the community which occasioned the run on the Bank, and it would equally have taken place if they had not made any advances to government and had possessed twice their present capital. If the Bank had continued paying in cash, probably the panic would have subsided before their coin had been exhausted.

"With the known opinion of the Bank directors as to the rule for issuing paper money, they may be said to have exercised their powers without any great indiscretion. It is evident that they have followed their own principle with extreme caution. In the present state of the law, they have the power, without any control whatever, of increasing or reducing the circulation in any degree they may think proper; a power which should neither be entrusted to the state itself, nor to anybody in it, as there can be no security for the uniformity in the value of the currency when its augmentation or diminution depends solely on the will of the issuers. That the Bank have the power of reducing the circulation to the very narrowest limits will not be denied, even by those who agree in opinion with the directors that they have not the power of adding indefinitely to its quantity. Though I am fully assured that it is both against the interest and the wish of the Bank to exercise this power to the detriment of the public, yet, when I contemplate the evil consequences which might ensue from a sudden and great reduction of the circulation, as well as from a great addition to it, I cannot but deprecate the facility with which the state has armed the Bank with so formidable a prerogative.

" The inconvenience to which country banks were subjected before the restriction on cash payments must at times have been very great. At all periods of alarm, or of expected alarm, they must have been under the necessity of providing themselves with guineas, that they might be prepared for every exigency which might occur. Guineas, on these occasions, were obtained at the Bank in exchange for the larger notes, and were conveyed by some confidential agent, at expense and risk, to the country bank. After performing the offices to which they were destined, they found their way again to London, and in all probability were again lodged in the Bank, provided they had not suffered such a loss of weight as to reduce them below the legal standard.

" If the plan now proposed of paying bank notes in bullion be adopted, it would be necessary either to extend the same privilege to country banks, or to make bank notes a legal tender, in which latter case there would be no alteration in the law respecting country banks, as they would be required, precisely as they now are, to pay their notes when demanded in Bank of England notes.

" The saving which would take place from not submitting the guineas to the loss of weight from the friction which they must undergo in their repeated journeys, as well as of the expenses of conveyance, would be considerable; but by far the greatest advantage would result from the permanent supply of the country as well as of the London circulation, as far as the smaller pay-ments are concerned, being provided in the very cheap medium paper, instead of the very valuable medium, gold; thereby enabling the country to derive all the profit which may be obtained by the productive employment of a capital to that amount. We should surely not be justified in rejecting so decided a benefit unless some specific inconvenience could be pointed out as likely to follow from adopting the cheaper medium."]

A currency is in its most perfect state when it consists wholly of paper money, but of paper money of an equal value with the gold which it professes to represent. The use of paper instead of gold substitutes the cheapest in place of the most expensive medium, and enables the country, without loss to any individual, to exchange all the gold which it before used for this purpose for raw materials, utensils, and food; by the use of which both its wealth and its enjoyments are increased.

In a national point of view, it is of no importance whether the issuers of this well regulated paper money be the government or a bank, it will, on the whole, be equally productive of riches whether it be issued by one or by the other; but it is not so with respect to the interest of individuals. In a country where the market rate of interest is 7 per cent., and where the state requires for a particular expense £70,000 per annum, it is a question of importance to the individuals of that country whether they must be taxed to pay this £70,000 per annum, or whether they could raise it without taxes. Suppose that a million of money should be required to fit out an expedition. If the state issued a million of paper and displaced a million of coin, the expedition would be fitted out without any charge to the people; but if a bank issued a million of paper, and lent it to government at 7 per cent., thereby displacing a million of coin, the country would be charged with a continual tax of

£70,000 per annum: the people would pay the tax, the bank would receive it, and the society would in either case be as wealthy as before; the expedition would have been really fitted out by the improvement of our system, by rendering capital of the value of a million productive in the form of commodities instead of letting it remain unproductive in the form of coin; but the advantage would always be in favour of the issuers of paper; and as the state represents the people, the people would have saved the tax if they, and not the bank, had issued this million.

I have already observed that if there were perfect security that the power of issuing paper money would not be abused, it would be of no importance with respect to the riches of the country collectively by whom it was issued; and I have now shown that the public would have a direct interest that the issuers should be the state, and not a company of merchants or bankers. The danger, however, is that this power would be more likely to be abused if in the hands of government than if in the hands of a banking company. A company would, it is said, be more under the control of law, and although it might be their interest to extend their issues beyond the bounds of discretion, they would be limited and checked by the power which individuals would have of calling for bullion or specie. It is argued that the same check would not be long respected if government had the privilege of issuing money; that they would be too apt to consider present convenience rather than future security, and might, therefore, on the alleged grounds of expediency, be too much inclined to remove the checks by which the amount of their issues was controlled.

Under an arbitrary government this objection would have great force; but in a free country, with an enlightened legislature, the power of issuing paper money, under the requisite checks of convertibility at the will of the holder, might be safely lodged in the hands of commissioners appointed for that special purpose, and they might be made totally independent of the control of ministers.

The sinking fund is managed by commissioners responsible only to Parliament, and the investment of the money entrusted to their charge proceeds with the utmost regularity; what reason can there be to doubt that the issues of paper money might be regulated with equal fidelity, if placed under similar management?

It may be said that although the advantage accruing to the

state, and, therefore, to the public, from issuing paper money is sufficiently manifest, as it would exchange a portion of the national debt, on which interest is paid by the public, into a debt bearing no interest: yet it would be disadvantageous to commerce, as it would preclude the merchants from borrowing money and getting their bills discounted, the method in which bank paper is partly issued.

This, however, is to suppose that money could not be borrowed if the Bank did not lend it, and that the market rate of interest and profit depends on the amount of the issues of money and on the channel through which it is issued. But as a country would have no deficiency of cloth, of wine, or any other commodity, if they had the means of paying for it, in the same manner neither would there be any deficiency of money to be lent if the borrowers offered good security and were willing to pay the market rate of interest for it.

In another part of this work I have endeavoured to show that the real value of a commodity is regulated, not by the accidental advantages which may be enjoyed by some of its producers, but by the real difficulties encountered by that producer who is least favoured. It is so with respect to the interest for money; it is not regulated by the rate at which the bank will lend, whether it be 5, 4, or 3 per cent., but by the rate of profits which can be made by the employment of capital, and which is totally independent of the quantity or of the value of money. Whether a bank lent one million, ten million, or a hundred millions, they would not permanently alter the market rate of interest; they would alter only the value of the money which they thus issued. In one case, ten or twenty times more money might be required to carry on the same business than what might be required in the other. The applications to the bank for money, then, depend on the comparison between the rate of profits that may be made by the employment of it and the rate at which they are willing to lend it. If they charge less than the market rate of interest, there is no amount of money which they might not lend; if they charge more than that rate none but spendthrifts and prodigals would be found to borrow of them. We accordingly find that when the market rate of interest exceeds the rate of 5 per cent. at which the Bank uniformly lend, the discount office is besieged with applicants for money; and, on the contrary, when the market rate is even temporarily under 5 per cent., the clerks of that office have no employment.

The reason, then, why for the last twenty years the Bank is said to have given so much aid to commerce, by assisting the merchants with money, is because they have, during that whole period, lent money below the market rate of interest; below that rate at which the merchants could have borrowed elsewhere; but I confess that to me this seems rather an objection to their establishment than an argument in favour of it.

What should we say of an establishment which should regularly supply half the clothiers with wool under the market price? Of what benefit would it be to the community? It would not extend our trade, because the wool would equally have been bought if they had charged the market price for it. It would not lower the price of cloth to the consumer, because the price, as I have said before, would be regulated by the cost of its production to those who were the least favoured. Its sole effect, then, would be to swell the profits of a part of the clothiers beyond the general and common rate of profits. The establishment would be deprived of its fair profits, and another part of the community would be in the same degree benefited. Now, this is precisely the effect of our banking establishments; a rate of interest is fixed by the law below that at which it can be borrowed in the market, and at this rate the Bank are required to lend or not to lend at all. From the nature of their establishment, they have large funds which they can only dispose of in this way; and a part of the traders of the country are unfairly, and, for the country, unprofitably benefited, by being enabled to supply themselves with an instrument of trade at a less charge than those who must be influenced only by a market price.

The whole business which the whole community can carry on depends on the quantity of its capital, that is, of its raw material, machinery, food, vessels, etc., employed in production. After a well regulated paper money is established, these can neither be increased nor diminished by the operations of banking. If, then, the state were to issue the paper money of the country, although it should never discount a bill, or lend one shilling to the public, there would be no alteration in the amount of trade; for we should have the same quantity of raw materials, of machinery, food, and ships; and it is probable, too, that the same amount of money might be lent, not always at 5 per cent., indeed, a rate fixed by law, when that might be under the market rate, but at 6, 7, or 8 per cent., the result of the fair competition in the market between the lenders and the borrowers.

Adam Smith speaks of the advantages derived by merchants from the superiority of the Scotch mode of affording accommodation to trade over the English mode, by means of cash accounts. These cash accounts are credits given by the Scotch banker to his customers, in addition to the bills which he discounts for them; but as the banker, in proportion as he advances money and sends it into circulation in one way, is debarred from issuing so much in the other, it is difficult to perceive in what the advantage consists. If the whole circulation will bear only one million of paper, one million only will be circulated; and it can be of no real importance either to the banker or merchant whether the whole be issued in discounting bills, or a part be so issued, and the remainder be issued by means of these cash accounts.

It may perhaps be necessary to say a few words on the subject of the two metals, gold and silver, which are employed in currency, particularly as this question appears to perplex, in many people's minds, the plain and simple principles of currency. " In England," says Dr. Smith, " gold was not considered as a legal tender for a long time after it was coined into money. The proportion between the values of gold and silver money was not fixed by any public law or proclamation, but was left to be settled by the market. If a debtor offered payment in gold, the creditor might either reject such payment altogether, or accept of it at such a valuation of the gold as he and his debtor could agree upon."

In this state of things it is evident that a guinea might sometimes pass for 22s. or more, and sometimes for 18s. or less, depending entirely on the alteration in the relative market value of gold and silver. All the variations, too, in the value of gold, as well as in the value of silver, would be rated in the gold coin—it would appear as if silver was invariable, and as if gold only was subject to rise and fall. Thus, although a guinea passed for 22s. instead of 18s., gold might not have varied in value; the variation might have been wholly confined to the silver, and therefore 22s. might have been of no more value than 18s. were before. And, on the contrary, the whole variation might have been in the gold; a guinea which was worth 18s. might have risen to the value of 22s.

If, now, we suppose this silver currency to be debased by clipping, and also increased in quantity, a guinea might pass for 30s. ; for the silver in 30s. of such debased money might be of no more value than the gold in one guinea. By restoring the silver currency to its Mint value, silver money would rise; but it would appear as if gold fell, for a guinea would probably be of no more value than 21 of such good shillings.

If now gold be also made a legal tender, and every debtor be at liberty to discharge a debt by the payment of 420 shillings, or twenty guineas for every £21 that he owes, he will pay in one or the other according as he can most cheaply discharge his debt. If with five quarters of wheat he can procure as much gold bullion as the Mint will coin into twenty guineas, and for the same wheat as much silver bullion as the Mint will coin for him into 430 shillings, he will prefer paying in silver, because he would be a gainer of ten shillings by so paying his debt. But if, on the contrary, he could obtain with this wheat as much gold as would be coined into twenty guineas and a half, and as much silver only as would coin into 420 shillings, he would naturally prefer paying his debt in gold. If the quantity of gold which he could procure could be coined only into twenty guineas, and the quantity of silver into 420 shillings, it would be a matter of perfect indifference to him in which money, silver or gold, it was that he paid his debt. It is not, then, a matter of chance; it is not because gold is better fitted for carrying on the circulation of a rich country that gold is ever preferred for the purpose of paying debts, but simply because it is the interest of the debtor so to pay them.

During a long period previous to 1797, the year of the restriction on the Bank payments in coin, gold was so cheap, compared with silver, that it suited the Bank of England, and all other debtors, to purchase gold in the market, and not silver, for the purpose of carrying it to the Mint to be coined, as they could in that coined metal more cheaply discharge their debts. The silver currency was, during a great part of this period, very much debased; but it existed in a degree of scarcity, and therefore, on the principle which I have before explained, it never sunk in its current value. Though so debased, it was still the interest of debtors to pay in the gold coin. If, indeed, the quantity of this debased silver coin had been enormously great, or if the Mint had issued such debased pieces, it might have been the interest of debtors to pay in this debased money; but its quantity was limited, and it sustained its value, and, therefore, gold was in practice the real standard of currency.

That it was so is nowhere denied; but it has been contended that it was made so by the law, which declared that silver should not be a legal tender for any debt exceeding £25, unless by weight, according to the Mint standard.

But this law did not prevent any debtor from paying his debt, however large its amount, in silver currency fresh from the Mint; that the debtor did not pay in this metal was not a matter of chance nor a matter of compulsion, but wholly the effect of choice; it did not suit him to take silver to the Mint, it did suit him to take gold thither. It is probable that if the quantity of this debased silver in circulation had been enormously great, and also a legal tender, that a guinea would have been again worth thirty shillings; but it would have been the debased shilling that would have fallen in value, and not the guinea that had risen.

It appears, then, that whilst each of the two metals was equally a legal tender for debts of any amount, we were subject to a constant change in the principal standard measure of value. It would sometimes be gold, sometimes silver, depending entirely on the variations in the relative value of the two metals; and at such times the metal which was not the standard would be melted and withdrawn from circulation, as its value would be greater in bullion than in coin. This was an inconvenience which it was highly desirable should be remedied; but so slow is the progress of improvement that, although it had been unanswerably demonstrated by Mr. Locke, and had been noticed by all writers on the subject of money since his day, a better system was never adopted till the session of Parliament 1816, when it was enacted that gold only should be a legal tender for any sum exceeding forty shillings.

Dr. Smith does not appear to have been quite aware of the effect of employing two metals as currency, and both a legal tender for debts of any amount; for he says that " in reality, during the continuance of any one regulated proportion between the respective values of the different metals in coin, the value of the most precious metal regulates the value of the whole coin." Because gold was in his day the medium in which it suited debtors to pay their debts, he thought that it had some inherent quality by which it did then, and always would, regulate the value of silver coin.

On the reformation of the gold coin in 1774, a new guinea fresh from the Mint would exchange for only twenty-one debased shillings; but in the reign of King William, when the silver coin was in precisely the same condition, a guinea also new and fresh from the Mint would exchange for thirty shillings. On

this Mr. Buchanan observes, " here, then, is a most singular
fact, of which the common theories of currency offer no account;
the guinea exchanging at one time for thirty shillings, its
intrinsic worth in a debased silver currency, and afterwards the
same guinea exchanged for only twenty-one of those debased
shillings. It is clear that some great change must have inter-
vened in the state of the currency between these two different
periods, of which Dr. Smith's hypothesis offers no explanation."

It appears to me that the difficulty may be very simply solved
by referring this different state of the value of the guinea at the
two periods mentioned to the different *quantities* of debased
silver currency in circulation. In King William's reign gold
was not a legal tender; it passed only at a conventional value.
All the large payments were probably made in silver, particularly as paper currency and the operations of banking were then
little understood. The quantity of this debased silver money
exceeded the quantity of silver money which would have been
maintained in circulation if nothing but undebased money had
been in use; and, consequently, it was depreciated as well as
debased. But in the succeeding period, when gold was a legal
tender, when bank notes also were used in effecting payments,
the quantity of debased silver money did not exceed the quantity
of silver coin fresh from the Mint which would have circulated
if there had been no debased silver money; hence, though the
money was debased it was not depreciated. Mr. Buchanan's
explanation is somewhat different; he thinks that a subsidiary
currency is not liable to depreciation, but that the main currency
is. In King William's reign silver was the main currency,
and hence was liable to depreciation. In 1774 it was a subsidiary currency, and, therefore, maintained its value. Depreciation, however, does not depend on a currency being the
subsidiary or the main currency, it depends wholly on its being
in excess of quantity.[1]

---

[1] It has lately been contended in Parliament by Lord Lauderdale that,
with the existing Mint regulation, the Bank could not pay their notes in
specie, because the relative value of the two metals is such that it would
be for the interest of all debtors to pay their debts with silver and not
with gold coin, while the law gives a power to all the creditors of the
Bank to demand gold in exchange for Bank notes. This gold, his lordship
thinks, could be profitably exported, and if so, he contends that the Bank,
to keep a supply, will be obliged to buy gold constantly at a premium
and sell it at par. If every other debtor could pay in silver, Lord Lauderdale would be right; but he cannot do so if his debt exceed 40s. This,
then, would limit the amount of silver coin in circulation (if government
had not reserved to itself the power to stop the coinage of that metal
whenever they might think it expedient); because if too much silver were

To a moderate seignorage on the coinage of money there cannot be much objection, particularly on that currency which is to effect the smaller payments. Money is generally enhanced in value to the full amount of the seignorage, and, therefore, it is a tax which in no way affects those who pay it, while the quantity of money is not in excess. It must, however, be remarked that in a country where a paper currency is established, although the issuers of such paper should be liable to pay it in specie on the demand of the holder, still, both their notes and the coin might be depreciated to the full amount of the seignorage on that coin, which is alone the legal tender, before the check, which limits the circulation of paper, would operate. If the seignorage of gold coin were 5 per cent. for instance, the currency, by an abundant issue of bank notes, might be really depreciated 5 per cent. before it would be the interest of the holders to demand coin for the purpose of melting it into bullion; a depreciation to which we should never be exposed if either there was no seignorage on the gold coin or, if a seignorage were allowed, the holders of bank notes might demand bullion, and not coin, in exchange for them, at the Mint price of £3 17s. 10½d. Unless, then, the Bank should be obliged to pay their notes in bullion or coin, at the will of the holder, the late law which allows a seignorage of 6 per cent., or fourpence per oz., on the silver coin, but which directs that gold shall be coined by the Mint without any charge whatever, is perhaps the most proper, as it will most effectually prevent any unnecessary variation of the currency.

coined it would sink in relative value to gold, and no man would accept it in payment for a debt exceeding 40s., unless a compensation were made for its lower value. To pay a debt of £100, 100 sovereigns, or bank notes to the amount of £100, would be necessary, but £105 in silver coin might be required if there were too much silver in circulation. There are, then, two checks against an excessive quantity of silver coin; first, the direct check which government may at any time interpose to prevent more from being coined; secondly, no motive of interest would lead any one to take silver to the Mint, if he might do so, for if it were coined, it would not pass current at its Mint but only at its market value

CHAPTER XXVIII

ON THE COMPARATIVE VALUE OF GOLD, CORN, AND LABOUR IN RICH AND POOR COUNTRIES

" Gold and silver, like all other commodities," says Adam Smith, " naturally seek the market where the best price is given for them; and the best price is commonly given for everything in the country which can best afford it. Labour, it must be remembered, is the ultimate price which is paid for everything; and in countries where labour is equally well rewarded, the money price of labour will be in proportion to that of the subsistence of the labourer. But gold and silver will naturally exchange for a greater quantity of subsistence in a rich than in a poor country; in a country which abounds with subsistence, than in one which is but indifferently supplied with it."

But corn is a commodity, as well as gold, silver, and other things; if all commodities, therefore, have a high exchangeable value in a rich country, corn must not be excepted; and hence we might correctly say that corn exchanged for a great deal of money because it was dear, and that money, too, exchanged for a great deal of corn because that also was dear; which is to assert that corn is dear and cheap at the same time. No point in political economy can be better established than that a rich country is prevented from increasing in population, in the same ratio as a poor country, by the progressive difficulty of providing food. That difficulty must necessarily raise the relative price of food and give encouragement to its importation. How then can money, or gold and silver, exchange for more corn in rich, than in poor, countries? It is only in rich countries, where corn is dear, that landholders induce the legislature to prohibit the importation of corn. Who ever heard of a law to prevent the importation of raw produce in America or Poland?—Nature has effectually precluded its importation by the comparative facility of its production in those countries.

How, then, can it be true that, " if you except corn, and such other vegetables as are raised altogether by human industry, all other sorts of rude produce—cattle, poultry, game of all kinds, the useful fossils and minerals of the earth, etc., naturally grow dearer as the society advances." Why should corn and vegetables alone be excepted? Dr. Smith's error, throughout his whole work, lies in supposing that the value of corn is constant; that though the value of all other things may, the value of corn never can, be raised. Corn, according to him, is always of the same value, because it will always feed the same number of people. In the same manner, it might be said that cloth is always of the same value, because it will always make the same number of coats. What can value have to do with the power of feeding and clothing?

Corn, like every other commodity, has in every country its natural price, viz. that price which is necessary to its production, and without which it could not be cultivated: it is this price which governs its market price, and which determines the expediency of exporting it to foreign countries. If the importation of corn were prohibited in England, its natural price might rise to £6 per quarter in England, whilst it was only at half that price in France. If at this time the prohibition of importation were removed, corn would fall in the English market, not to a price between £6 and £3, but ultimately and permanently to the natural price of France, the price at which it could be furnished to the English market and afford the usual and ordinary profits of stock in France; and it would remain at this price whether England consumed a hundred thousand or a million of quarters. If the demand of England were for the latter quantity, it is probable that, owing to the necessity under which France would be of having recourse to land of a worse quality, to furnish this large supply, the natural price would rise in France; and this would of course affect also the price of corn in England. All that I contend for is, that it is the natural price of commodities in the exporting country which ultimately regulates the prices at which they shall be sold, if they are not the objects of monopoly in the importing country.

But Dr. Smith, who has so ably supported the doctrine of the natural price of commodities ultimately regulating their market price, has supposed a case in which he thinks that the market price would not be regulated either by the natural price of the exporting or of the importing country. " Diminish the real opulence either of Holland or the territory of Genoa," he says, " while the number of their inhabitants remains the same; diminish their power of supplying themselves from distant countries, and the price of corn, instead of sinking with that diminution in the quantity of their silver which must necessarily accompany this declension, either as its cause or as its effect, will rise to the price of a famine."

To me it appears that the very reverse would take place: the diminished power of the Dutch or Genoese to purchase generally might depress the price of corn for a time below its natural price in the country from which it was exported, as well as in the countries in which it was imported; but it is quite impossible that it could ever raise it above that price. It is only by increasing the opulence of the Dutch and Genoese that you could increase the demand, and raise the price of corn above its former price; and that would take place only for a very limited time, unless new difficulties should arise in obtaining the supply.

Dr. Smith further observes on this subject: " When we are in want of necessaries we must part with all superfluities, of which the value, as it rises in times of opulence and prosperity, so it sinks in times of poverty and distress." This is undoubtedly true; but he continues, " it is otherwise with necessaries. Their real price, the quantity of labour which they can purchase or command, rises in times of poverty and distress, and sinks in times of opulence and prosperity, which are always times of great abundance, for they could not otherwise be times of opulence and prosperity. Corn is a necessary, silver is only a superfluity."

Two propositions are here advanced which have no connection with each other; one, that under the circumstances supposed, corn would command more labour, which is not disputed; the other, that corn would sell at a higher money price, that it would exchange for more silver; this I contend to be erroneous. It might be true if corn were at the same time scarce—if the usual supply had not been furnished. But in this case it is abundant; it is not pretended that a less quantity than usual is imported, or that more is required. To purchase corn, the Dutch or Genoese want money, and to obtain this money they are obliged to sell their superfluities. It is the market value and price of these superfluities which falls, and money appears to rise as compared with them. But this will not tend to increase the demand for corn, nor to lower the value of money, the only two causes which can raise the price of corn. Money, from a want of credit, and from other causes, may be in great demand, and consequently dear, comparatively with corn; but on no just principle can it be maintained that under such circumstances money would be cheap and, therefore, that the price of corn would rise.

When we speak of the high or low value of gold, silver, or any other commodity in different countries, we should always mention some medium in which we are estimating them, or no idea can be attached to the proposition. Thus, when gold is said to be dearer in England than in Spain, if no commodity is mentioned, what notion does the assertion convey? If corn, olives, oil, wine, and wool be at a cheaper price in Spain than in England, estimated in those commodities gold is dearer in Spain. If, again, hardware, sugar, cloth, etc., be at a lower price in England than in Spain, then, estimated in those commodities, gold is dearer in England. Thus gold appears dearer or cheaper in Spain as the fancy of the observer may fix on the medium by which he estimates its value. Adam Smith, having stamped corn and labour as a universal measure of value, would naturally estimate the comparative value of gold by the quantity of those two objects for which it would exchange: and, accordingly, when he speaks of the comparative value of gold in two countries, I understand him to mean its value estimated in corn and labour.

But we have seen that, estimated in corn, gold may be of very different value in two countries. I have endeavoured to show that it will be low in rich countries and high in poor countries; Adam Smith is of a different opinion: he thinks that the value of gold, estimated in corn, is highest in rich countries. But without further examining which of these opinions is correct, either of them is sufficient to show that gold will not necessarily be lower in those countries which are in possession of the mines, though this is a proposition maintained by Adam Smith. Suppose England to be possessed of the mines, and Adam Smith's opinion, that gold is of the greatest value in rich countries, to be correct: although gold would naturally flow from England to all other countries in exchange for their *goods*, it would not follow that gold was necessarily lower in England, as compared with corn and labour, than in those countries. In another place, however, Adam Smith speaks of the precious metals being necessarily lower in Spain and Portugal than in other parts of Europe, because those countries happen to be almost the exclusive possessors of the mines which produce them. " Poland, where the feudal system still continues to take place, is at this day as beggarly a country as it was before the discovery of America. *The money price of corn, however, has risen;* THE REAL VALUE OF THE PRECIOUS METALS HAS FALLEN in Poland in the same manner as in other parts of Europe. Their quantity, therefore, must have increased there as in other places, *and nearly in the same proportion to the annual produce of the land and labour.* This increase of the quantity of those metals, however, has not, it seems, increased that annual produce; has neither improved the manufactures and agriculture of the country, nor mended the circumstances of its inhabitants. Spain and Portugal, the countries which possess the mines, are, after Poland, perhaps the two most beggarly countries in Europe. The value of the precious metals, however, *must be lower in Spain and Portugal* than in any other parts of Europe, loaded not only with a freight and insurance, but with the expense of smuggling, their exportation being either prohibited or subjected to a duty. *In proportion to the annual produce of the land and labour, therefore, their quantity must be greater in* those countries than in any other part of Europe: those countries, however, are poorer than the greater part of Europe. Though the feudal system has been abolished in Spain and Portugal, it has not been succeeded by a much better."

Dr. Smith's argument appears to me to be this: Gold, when estimated in corn, is cheaper in Spain than in other countries, and the proof of this is not that corn is given by other countries to Spain for gold, but that cloth, sugar, hardware, are by those countries given in exchange for that metal.

CHAPTER XXIX

TAXES PAID BY THE PRODUCER

M. SAY greatly magnifies the inconveniences which result if a tax on a manufactured commodity is levied at an early, rather than at a late, period of its manufacture. The manufacturers, he observes, through whose hands the commodity may successively pass, must employ greater funds in consequence of having to advance the tax, which is often attended with considerable difficulty to a manufacturer of very limited capital and credit. To this observation no objection can be made.

Another inconvenience on which he dwells is that, in consequence of the advance of the tax, the profits on the advance also must be charged to the consumer, and that this additional tax is one from which the treasury derives no advantage.

In this latter objection I cannot agree with M. Say. The state, we will suppose, wants to raise *immediately* £1000, and levies it on a manufacturer, who will not for a twelvemonth be able to charge it to the consumer on his finished commodity. In consequence of such delay, he is obliged to charge for his commodity an additional price, not only of £1000, the amount of the tax, but probably of £1100, £100 being for interest on the £1000 advanced. But in return for this additional £100 paid by the consumer, he has a real benefit, inasmuch as his payment of the tax which government required immediately, and which he must finally pay, has been postponed for a year; an opportunity, therefore, has been afforded to him of lending to the manufacturer who had occasion for it the £1000, at 10 per cent., or at any other rate of interest which might be agreed upon. Eleven hundred pounds, payable at the end of one year, when money is at 10 per cent. interest, is of no more value than £1000 to be paid immediately. If government delayed receiving the tax for one year till the manufacture of the commodity was completed, it would perhaps be obliged to issue an exchequer bill bearing interest, and it would pay as much for interest as the consumer would save in price, excepting, indeed, that portion of the price which the manufacturer might be enabled, in consequence of the tax, to add to his own real gains. If for

258

the interest of the exchequer bill government would pay 5 per cent., a tax of £50 is saved by not issuing it. If the manufacturer borrowed the additional capital at 5 per cent., and charged the consumer 10 per cent., he also will have gained 5 per cent. on his advance, over and above his usual profits, so that the manufacturer and government together gain or save precisely the sum which the consumer pays.

M. Simonde, in his excellent work, *De la Richesse Commerciale*, following the same line of argument as M. Say, has calculated that a tax of 4000 francs, paid originally by a manufacturer, whose profits were at the moderate rate of 10 per cent., would, if the commodity manufactured only passed through the hands of five different persons, be raised to the consumer to the sum of 6734 francs. This calculation proceeds on the supposition that he who first advanced the tax would receive from the next manufacturer 4400 francs, and he again from the next, 4840 francs; so that at each step 10 per cent. on its value would be added to it. This is to suppose that the value of the tax would be accumulating at compound interest; not at the rate of 10 per cent. per annum, but at an absolute rate of 10 per cent. at every step of its progress. This opinion of M. de Simonde would be correct if five years elapsed between the first advance of the tax and the sale of the taxed commodity to the consumer; but if one year only elapsed, a remuneration of 400 francs, instead of 2734, would give a profit at the rate of 10 per cent. per annum to all who had contributed to the advance of the tax, whether the commodity had passed through the hands of five manufacturers or fifty.

CHAPTER XXX

ON THE INFLUENCE OF DEMAND AND SUPPLY ON PRICES

It is the cost of production which must ultimately regulate the price of commodities, and not, as has been often said, the proportion between the supply and demand: the proportion between supply and demand may, indeed, for a time, affect the market value of a commodity, until it is supplied in greater or less abundance, according as the demand may have increased or diminished; but this effect will be only of temporary duration.

Diminish the cost of production of hats, and their price will ultimately fall to their new natural price, although the demand should be doubled, trebled, or quadrupled. Diminish the cost of subsistence of men, by diminishing the natural price of the food and clothing by which life is sustained, and wages will ultimately fall, notwithstanding that the demand for labourers may very greatly increase.

The opinion that the price of commodities depends solely on the proportion of supply to demand, or demand to supply, has become almost an axiom in political economy, and has been the source of much error in that science. It is this opinion which has made Mr. Buchanan maintain that wages are not influenced by a rise or fall in the price of provisions, but solely by the demand and supply of labour; and that a tax on the wages of labour would not raise wages, because it would not alter the proportion of the demand of labourers to the supply.

The demand for a commodity cannot be said to increase if no additional quantity of it be purchased or consumed; and yet under such circumstances its money value may rise. Thus, if the value of money were to fall, the price of every commodity would rise, for each of the competitors would be willing to spend more money than before on its purchase; but though its price rose 10 or 20 per cent., if no more were bought than before, it would not, I apprehend, be admissible to say that the variation in the price of the commodity was caused by the increased demand for it. Its natural price, its money cost of production, would be really altered by the altered value of money; and without any increase of demand, the price of the commodity would be naturally adjusted to that new value.

" We have seen," says M. Say, " that the cost of production determines the lowest price to which things can fall: the price below which they cannot remain for any length of time, because production would then be either entirely stopped or diminished." Vol. ii. p. 26.

He afterwards says that the demand for gold having increased in a still greater proportion than the supply, since the discovery of the mines, " its price in goods, instead of falling in the proportion of ten to one, fell only in the proportion of four to one; " that is to say, instead of falling in proportion as its natural price had fallen, fell in proportion as the supply exceeded the demand.[1]—" *The value of every commodity rises always in a direct ratio to the demand, and in an inverse ratio to the supply.*"

The same opinion is expressed by the Earl of Lauderdale.

" With respect to the variations in value, of which everything valuable is susceptible, if we could for a moment suppose that any substance possessed intrinsic and fixed value, so as to render an assumed quantity of it constantly, under all circumstances, of an equal value, then the degree of value of all things, ascertained by such a fixed standard, would vary according to the proportion *betwixt the quantity of them* and the demand for them, and every commodity would, of course, be subject to a variation in its value, from four different circumstances:

1. " It would be subject to an increase of its value, from a diminution of its quantity.

2. " To a diminution of its value, from an augmentation of its quantity.

3. " It might suffer an augmentation in its value, from the circumstance of an increased demand.

4. " Its value might be diminished by a failure of demand.

" As it will, however, clearly appear that no commodity can possess fixed and intrinsic value, so as to qualify it for a measure of the value of other commodities, mankind are induced to select, as a practical measure of value, that which appears the least

---

[1] If, with the quantity of gold and silver which actually exists, these metals only served for the manufacture of utensils and ornaments, they would be abundant, and would be much cheaper than they are at present: in other words, in exchanging them for any other species of goods, we should be obliged to give proportionally a greater quantity of them. But as a large quantity of these metals is used for money, and as this portion is used for no other purpose, there remains less to be employed in furniture and jewellery; now this scarcity adds to their value.—Say, vol. ii. p. 316.

liable to any of these four sources of variations, *which are the sole causes of alteration of value.*

"When, in common language, therefore, we express the *value* of any commodity, it may vary at one period from what it is at another, in consequence of eight different contingencies:—

1. "From the four circumstances above stated, in relation to the commodity of which we mean to express the value.

2. "From the same four circumstances, in relation to the commodity we have adopted as a measure of value." [1]

This is true of monopolised commodities, and, indeed, of the market price of all other commodities for a limited period. If the demand for hats should be doubled, the price would immediately rise, but that rise would be only temporary, unless the cost of production of hats or their natural price were raised. If the natural price of bread should fall 50 per cent. from some great discovery in the science of agriculture, the demand would not greatly increase, for no man would desire more than would satisfy his wants, and as the demand would not increase, neither would the supply; for a commodity is not supplied merely because it can be produced, but because there is a demand for it. Here, then, we have a case where the supply and demand have scarcely varied, or, if they have increased, they have increased in the same proportion; and yet the price of bread will have fallen 50 per cent., at a time, too, when the value of money had continued invariable.

Commodities which are monopolised, either by an individual or by a company, vary according to the law which Lord Lauderdale has laid down: they fall in proportion as the sellers augment their quantity, and rise in proportion to the eagerness of the buyers to purchase them; their price has no necessary connection with their natural value: but the prices of commodities which are subject to competition, and whose quantity may be increased in any moderate degree, will ultimately depend, not on the state of demand and supply, but on the increased or diminished cost of their production.

[1] *An Inquiry into the Nature and Origin of Public Wealth*, p. 13.

CHAPTER XXXI

ON MACHINERY

In the present chapter I shall enter into some inquiry respecting the influence of machinery on the interests of the different classes of society, a subject of great importance, and one which appears never to have been investigated in a manner to lead to any certain or satisfactory results. It is more incumbent on me to declare my opinion on this question, because they have, on further reflection, undergone a considerable change; and although I am not aware that I have ever published anything respecting machinery which it is necessary for me to retract, yet I have in other ways given my support to doctrines which I now think erroneous; it therefore becomes a duty in me to submit my present views to examination, with my reasons for entertaining them.

Ever since I first turned my attention to questions of political economy, I have been of opinion that such an application of machinery to any branch of production as should have the effect of saving labour was a general good, accompanied only with that portion of inconvenience which in most cases attends the removal of capital and labour from one employment to another. It appeared to me that, provided the landlords had the same money rents, they would be benefited by the reduction in the prices of some of the commodities on which those rents were expended, and which reduction of price could not fail to be the consequence of the employment of machinery. The capitalist, I thought, was eventually benefited precisely in the same manner. He, indeed, who made the discovery of the machine, or who first usefully applied it, would enjoy an additional advantage by making great profits for a time; but, in proportion as the machine came into general use, the price of the commodity produced would, from the effects of competition, sink to its cost of production, when the capitalist would get the same money profits as before, and he would only participate in the general advantage as a consumer, by being enabled, with the same money revenue, to command an additional quantity of comforts and enjoyments. The class of labourers also, I thought, was equally benefited by the use of machinery, as they would have the means of buying more commodities with the same money wages, and I thought that no reduction of wages would take place because the capitalist would have the power of demanding and employing the same quantity of labour as before, although he might be under the necessity of employing it in the production of a new or, at any rate, of a different commodity. If, by improved machinery, with the employment of the same quantity of labour, the quantity of stockings could be quadrupled, and the demand for stockings were only doubled, some labourers would necessarily be discharged from the stocking trade; but as the capital which employed them was still in being, and as it was the interest of those who had it to employ it productively, it appeared to me that it would be employed on the production of some other commodity useful to the society, for which there could not fail to be a demand; for I was, and am, deeply impressed with the truth of the observation of Adam Smith, that " the desire for food is limited in every man by the narrow capacity of the human stomach, but the desire of the conveniences and ornaments of building, dress, equipage, and household furniture, seems to have no limit or certain boundary." As, then, it appeared to me that there would be the same demand for labour as before, and that wages would be no lower, I thought that the labouring class would, equally with the other classes, participate in the advantage, from the general cheapness of commodities arising from the use of machinery.

These were my opinions, and they continue unaltered, as far as regards the landlord and the capitalist; but I am convinced that the substitution of machinery for human labour is often very injurious to the interests of the class of labourers.

My mistake arose from the supposition that whenever the net income of a society increased, its gross income would also increase; I now, however, see reason to be satisfied that the one fund, from which landlords and capitalists derive their revenue, may increase, while the other, that upon which the labouring class mainly depend, may diminish, and therefore it follows, if I am right, that the same cause which may increase the net revenue of the country may at the same time render the population redundant, and deteriorate the condition of the labourer.

A capitalist, we will suppose, employs a capital of the value of £20,000, and that he carries on the joint business of a farmer and a manufacturer of necessaries. We will further suppose that £7000 of this capital is invested in fixed capital, viz. in buildings, implements, etc., etc., and that the remaining £13,000 is employed as circulating capital in the support of labour. Let us suppose, too, that profits are 10 per cent., and consequently that the capitalist's capital is every year put into its original state of efficiency and yields a profit of £2000.

Each year the capitalist begins his operations by having food and necessaries in his possession of the value of £13,000, all of which he sells in the course of the year to his own workmen for that sum of money, and, during the same period, he pays them the like amount of money for wages: at the end of the year they replace in his possession food and necessaries of the value of £15,000, £2000 of which he consumes himself, or disposes of as may best suit his pleasure and gratification. As far as these products are concerned, the gross produce for that year is £15,000, and the net produce £2000. Suppose, now, that the following year the capitalist employs half his men in constructing a machine, and the other half in producing food and necessaries as usual. During that year he would pay the sum of £13,000 in wages as usual, and would sell food and necessaries to the same amount to his workmen; but what would be the case the following year?

While the machine was being made, only one-half of the usual quantity of food and necessaries would be obtained, and they would be only one-half the value of the quantity which was produced before. The machine would be worth £7500, and the food and necessaries £7500, and, therefore, the capital of the capitalist would be as great as before; for he would have, besides these two values, his fixed capital worth £7000, making in the whole £20,000 capital, and £2000 profit. After deducting this latter sum for his own expenses, he would have a no greater circulating capital than £5500 with which to carry on his subsequent operations; and, therefore, his means of employing labour would be reduced in the proportion of £13,000 to £5500, and, consequently, all the labour which was before employed by £7500 would become redundant.

The reduced quantity of labour which the capitalist can employ, must, indeed, with the assistance of the machine, and after deductions for its repairs, produce a value equal to £7500, it must replace the circulating capital with a profit of £2000 on the whole capital; but if this be done, if the net income be not diminished, of what importance is it to the capitalist whether the gross income be of the value of £3000, of £10,000, or of £15,000?

In this case, then, although the net produce will not be diminished in value, although its power of purchasing commodities may be greatly increased, the gross produce will have fallen from a value of £15,000 to a value of £7500; and as the power of supporting a population, and employing labour, depends always on the gross produce of a nation, and not on its net produce, there will necessarily be a diminution in the demand for labour, population will become redundant, and the situation of the labouring classes will be that of distress and poverty.

As, however, the power of saving from revenue to add to capital must depend on the efficiency of the net revenue, to satisfy the wants of the capitalist, it could not fail to follow from the reduction in the price of commodities consequent on the introduction of machinery that with the same wants he would have increased means of saving—increased facility of transferring revenue into capital. But with every increase of capital he would employ more labourers; and, therefore, a portion of the people thrown out of work in the first instance would be subsequently employed; and if the increased production, in consequence of the employment of the machine, was so great as to afford, in the shape of net produce, as great a quantity of food and necessaries as existed before in the form of gross produce, there would be the same ability to employ the whole population, and, therefore, there would not necessarily be any redundancy of people.

All I wish to prove is that the discovery and use of machinery may be attended with a diminution of gross produce; and whenever that is the case, it will be injurious to the labouring class, as some of their number will be thrown out of employment, and population will become redundant compared with the funds which are to employ it.

The case which I have supposed is the most simple that I could select; but it would make no difference in the result if we supposed that the machinery was applied to the trade of any manufacturer — that of a clothier, for example, or of a cotton manufacturer. If, in the trade of a clothier, less cloth would be produced after the introduction of machinery, for a part of that quantity which is disposed of for the purpose of paying a large body of workmen would not be required by their employer. In consequence of using the machine, it would be necessary for him to reproduce a value only equal to the value consumed, together with the profits on the whole capital.

£7500 might do this as effectually as £15,000 did before, the case differing in no respect from the former instance. It may be said, however, that the demand for cloth would be as great as before, and it may be asked from whence would this supply come? But by whom would the cloth be demanded? By the farmers and the other producers of necessaries, who employed their capitals in producing these necessaries as a means of obtaining cloth: they gave corn and necessaries to the clothier for cloth, and he bestowed them on his workmen for the cloth which their work afforded him.

This trade would now cease; the clothier would not want the food and clothing, having fewer men to employ and having less cloth to dispose of. The farmers and others, who only produced necessaries as means to an end, could no longer obtain cloth by such an application of their capitals, and, therefore, they would either themselves employ their capitals in producing cloth, or would lend them to others, in order that the commodity really wanted might be furnished; and that for which no one had the means of paying, or for which there was no demand, might cease to be produced. This, then, leads us to the same result; the demand for labour would diminish, and the commodities necessary to the support of labour would not be produced in the same abundance.

If these views be correct, it follows, first, that the discovery and useful application of machinery always leads to the increase of the net produce of the country, although it may not, and will not, after an inconsiderable interval, increase the value of that net produce.

Secondly, that an increase of the net produce of a country is compatible with a diminution of the gross produce, and that the motives for employing machinery are always sufficient to ensure its employment if it will increase the net produce, although it may, and frequently must, diminish both the quantity of the gross produce and its value.

Thirdly, that the opinion entertained by the labouring class, that the employment of machinery is frequently detrimental to their interests, is not founded on prejudice and error, but is conformable to the correct principles of political economy.

Fourthly, that if the improved means of production, in consequence of the use of machinery, should increase the net produce of a country in a degree so great as not to diminish the gross produce (I mean always quantity of commodities, and not value), then the situation of all classes will be improved. The landlord and capitalist will benefit, not by an increase of rent and profit, but by the advantages resulting from the expenditure of the same rent and profit on commodities very considerably reduced in value, while the situation of the labouring classes will also be considerably improved; First, from the increased demand for menial servants; secondly, from the stimulus to savings from revenue which such an abundant net produce will afford; and, thirdly, from the low price of all articles of consumption on which their wages will be expended.

Independently of the consideration of the discovery and use of machinery, to which our attention has been just directed, the labouring class have no small interest in the manner in which the net income of the country is expended, although it should, in all cases, be expended for the gratification and enjoyments of those who are fairly entitled to it.

If a landlord, or a capitalist, expends his revenue in the manner of an ancient baron, in the support of a great number of retainers, or menial servants, he will give employment to much more labour than if he expended it on fine clothes or costly furniture, on carriages, on horses, or in the purchase of any other luxuries.

In both cases the net revenue would be the same, and so would be the gross revenue, but the former would be realised in different commodities. If my revenue were £10,000, the same quantity nearly of productive labour would be employed whether I realised it in fine clothes and costly furniture, etc., etc., or in a quantity of food and clothing of the same value. If, however, I realised my revenue in the first set of commodities, no more labour would be *consequently* employed: I should enjoy my furniture and my clothes, and there would be an end of them; but if I realised my revenue in food and clothing, and my desire was to employ menial servants, all those whom I could so employ with my revenue of £10,000, or with the food and clothing which it would purchase, would be to be added to the former demand for labourers, and this addition would take place only because I chose this mode of expending my revenue. As the labourers, then, are interested in the demand for labour, they must naturally desire that as much of the revenue as possible should be diverted from expenditure on luxuries to be expended in the support of menial servants.

In the same manner, a country engaged in war, and which is under the necessity of maintaining large fleets and armies, employs a great many more men than will be employed when the war terminates, and the annual expenses which it brings with it, cease.

If I were not called upon for a tax of £500 during the war, and which is expended on men in the situations of soldiers and sailors, I might probably expend that portion of my income on furniture, clothes, books, etc., etc., and whether it was expended in the one way or in the other, there would be the same quantity of labour employed in production; for the food and clothing of the soldier and sailor would require the same amount of industry to produce it as the more luxurious commodities; but in the case of the war, there would be the additional demand for men as soldiers and sailors; and, consequently, a war which is supported out of the revenue, and not from the capital of a country, is favourable to the increase of population.

At the termination of the war, when part of my revenue reverts to me, and is employed as before in the purchase of wine, furniture, or other luxuries, the population which it before supported, and which the war called into existence, will become redundant, and by its effect on the rest of the population, and its competition with it for employment, will sink the value of wages, and very materially deteriorate the condition of the labouring classes.

There is one other case that should be noticed of the possibility of an increase in the amount of the net revenue of a country, and even of its gross revenue, with a diminution of demand for labour, and that is when the labour of horses is substituted for that of man. If I employed one hundred men on my farm, and if I found that the food bestowed on fifty of those men could be diverted to the support of horses, and afford me a greater return of raw produce, after allowing for the interest of the capital which the purchase of the horses would absorb, it would be advantageous to me to substitute the horses for the men, and I should accordingly do so; but this would not be for the interest of the men, and unless the income I obtained was so much increased as to enable me to employ the men as well as the horses, it is evident that the population would become redundant and the labourer's condition would sink in the general scale. It is evident he could not, under any circumstances, be employed in agriculture; but if the produce of the land were increased by the substitution of horses for men, he might be employed in manufactures, or as a menial servant.

The statements which I have made will not, I hope, lead to the inference that machinery should not be encouraged. To elucidate the principle, I have been supposing that improved machinery is *suddenly* discovered and extensively used; but the truth is that these discoveries are gradual, and rather operate in determining the employment of the capital which is saved and accumulated than in diverting capital from its actual employment.

With every increase of capital and population food will generally rise, on account of its being more difficult to produce. The consequence of a rise of food will be a rise of wages, and every rise of wages will have a tendency to determine the saved capital in a greater proportion than before to the employment of machinery. Machinery and labour are in constant competition, and the former can frequently not be employed until labour rises.

In America and many other countries, where the food of man is easily provided, there is not nearly such great temptation to employ machinery as in England, where food is high and costs much labour for its production. The same cause that raises labour does not raise the value of machines, and, therefore, with every augmentation of capital, a greater proportion of it is employed on machinery. The demand for labour will continue to increase with an increase of capital, but not in proportion to its increase; the ratio will necessarily be a diminishing ratio.[1]

I have before observed, too, that the increase of net incomes, estimated in commodities, which is always the consequence of improved machinery, will lead to new savings and accumula-

---

[1] " The demand for labour depends on the increasing of circulating and not of fixed capital. Were it true that the proportion between these two sorts of capital is the same at all times, and in all countries, then, indeed, it follows that the number of labourers employed is in proportion to the wealth of the state. But such a position has not the semblance of probability. As arts are cultivated, and civilisation is extended, fixed capital bears a larger and larger proportion to circulating capital. The amount of fixed capital employed in the production of a piece of British muslin is at least a hundred, probably a thousand times greater than that employed in the production of a similar piece of Indian muslin. And the proportion of circulating capital employed is a hundred or a thousand times less. It is easy to conceive that, under certain circumstances, the whole of the annual savings of an industrious people might be added to fixed capital, in which case they would have no effect in increasing the demand for labour."—Barton, *On the Condition of the Labouring Classes of Society*, page 16.

It is not easy, I think, to conceive that, under any circumstances, an increase of capital should not be followed by an increased demand for labour; the most that can be said is, that the demand will be in a diminishing ratio. Mr. Barton, in the above publication, has, I think, taken a correct view of some of the effects of an increasing amount of fixed capital on the condition of the labouring classes. His essay contains much valuable information.

tions. These savings, it must be remembered, are annual, and must soon create a fund much greater than the gross revenue originally lost by the discovery of the machine, when the demand for labour will be as great as before, and the situation of the people will be still further improved by the increased savings which the increased net revenue will still enable them to make.

The employment of machinery could never be safely discouraged in a state, for if a capital is not allowed to get the greatest net revenue that the use of machinery will afford here, it will be carried abroad, and this must be a much more serious discouragement to the demand for labour than the most extensive employment of machinery; for while a capital is employed in this country it must create a demand for some labour; machinery cannot be worked without the assistance of men, it cannot be made but with the contribution of their labour. By investing part of a capital in improved machinery there will be a diminution in the progressive demand for labour; by exporting it to another country the demand will be wholly annihilated.

The prices of commodities, too, are regulated by their cost of production. By employing improved machinery, the cost of production of commodities is reduced, and, consequently, you can afford to sell them in foreign markets at a cheaper price. If, however, you were to reject the use of machinery, while all other countries encouraged it, you would be obliged to export your money, in exchange for foreign goods, till you sunk the natural prices of your goods to the prices of other countries. In making your exchanges with those countries you might give a commodity which cost two days' labour here for a commodity which cost one abroad, and this disadvantageous exchange would be the consequence of your own act, for the commody which you export, and which cost you two days' labour, would have cost you only one if you had not rejected the use of machinery, the services of which your neighbours had more wisely appropriated to themselves.

CHAPTER XXXII

ALTHOUGH the nature of rent has in the former pages of this work been treated on at some length, yet I consider myself bound to notice some opinions on the subject which appear to me erroneous, and which are the more important as they are found in the writings of one to whom, of all men of the present day, some branches of economical science are the most indebted. Of Mr. Malthus's *Essay on Population* I am happy in the opportunity here afforded me of expressing my admiration. The assaults of the opponents of this great work have only served to prove its strength; and I am persuaded that its just reputation will spread with the cultivation of that science of which it is so eminent an ornament. Mr. Malthus, too, has satisfactorily explained the principles of rent, and showed that it rises or falls in proportion to the relative advantages, either of fertility or situation, of the different lands in cultivation, and has thereby thrown much light on many difficult points connected with the subject of rent, which were before either unknown or very imperfectly understood; yet he appears to me to have fallen into some errors which his authority makes it the more necessary, whilst his characteristic candour renders it less unpleasing, to notice. One of these errors lies in supposing rent to be a clear gain and a new creation of riches.

I do not assent to all the opinions of Mr. Buchanan concerning rent; but with those expressed in the following passage, quoted from his work by Mr. Malthus, I fully agree, and therefore I must dissent from Mr. Malthus's comment on them.

" In this view it (rent) can form no general addition to the stock of the community, as the neat surplus in question is nothing more than a revenue transferred from one class to another; and from the mere circumstance of its thus changing hands, it is clear that no fund can arise out of which to pay taxes. The revenue which pays for the produce of the land exists already in the hands of those who purchase that produce; and if the price of subsistence were lower, it would still remain in their hands, where it would be just as available for taxation as when, by a higher price, it is transferred to the landed proprietor."

After various observations on the difference between raw produce and manufactured commodities, Mr. Malthus asks, " Is it possible, then, with M. de Sismondi, to regard rent as the sole produce of labour, which has a value purely nominal, and the mere result of that augmentation of price which a seller obtains in consequence of a peculiar privilege; or, with Mr. Buchanan, to consider it as no addition to the national wealth, but merely a transfer of value, advantageous only to the landlords, and proportionably *injurious* to the consumers? " [1]

I have already expressed my opinion on this subject in treating of rent, and have now only further to add, that rent is a creation of value, as I understand that word, but not a creation of wealth. If the price of corn, from the difficulty of producing any portion of it, should rise from £4 to £5 per quarter, a million of quarters will be of the value of £5,000,000 instead of £4,000,000, and as this corn will exchange not only for more money, but for more of every other commodity, the possessors will have a greater amount of value; and as no one else will, in consequence, have a less, the society altogether will be possessed of greater value, and, in that sense, rent is a creation of value. But this value is so far nominal that it adds nothing to the wealth, that is to say, the necessaries, conveniences, and enjoyments of the society. We should have precisely the same quantity and no more of commodities, and the same million quarters of corn as before; but the effect of its being rated at £5 per quarter instead of £4 would be to transfer a portion of the value of the corn and commodities from their former possessors to the landlords. Rent, then, is a creation of value, but not a creation of wealth; it adds nothing to the resources of a country; it does not enable it to maintain fleets and armies; for the country would have a greater disposable fund if its land were of a better quality, and it could employ the same capital without generating a rent.

It must then be admitted that Mr. Sismondi and Mr. Buchanan, for both their opinions are substantially the same, were correct when they considered rent as a value purely nominal, and as forming no addition to the national wealth, but merely as a transfer of value, advantageous only to the landlords and proportionably injurious to the consumer.

In another part of Mr. Malthus's *Inquiry* he observes, " that the immediate cause of rent is obviously the excess of

[1] *An Inquiry into the Nature and Progress of Rent*, p. 15.

price above the cost of production at which raw produce sells in the market;" and, in another place, he says, "that the causes of the high price of raw produce may be stated to be three:—

"First, and mainly, that quality of the earth by which it can be made to yield a greater portion of the necessaries of life than is required for the maintenance of the persons employed on the land.

"Secondly, that quality peculiar to the necessaries of life, of being able to create their own demand, or to raise up a number of demanders in proportion to the quantity of necessaries produced.

"And thirdly, the comparative scarcity of the most fertile land." In speaking of the high price of corn, Mr. Malthus evidently does not mean the price per quarter or per bushel, but rather the excess of price for which the whole produce will sell above the cost of its production, including always in the term "cost of its production" profits as well as wages. One hundred and fifty quarters of corn at £3 10s. per quarter would yield a larger rent to the landlord than 100 quarters at £4, provided the cost of production were in both cases the same.

High price, if the expression be used in this sense, cannot then be called a *cause* of rent; it cannot be said "that the immediate cause of rent is obviously the excess of price above the cost of production, at which raw produce sells in the market," for that excess is itself rent. Rent Mr. Malthus has defined to be "that portion of the value of the whole produce which remains to the owner of the land after all the outgoings belonging to its cultivation, of whatever kind, have been paid, including the profits of the capital employed, estimated according to the usual and ordinary rate of the profits of agricultural stock at the time being." Now, whatever sum this excess may sell for, is money rent; it is what Mr. Malthus means by "the excess of price above the cost of production at which raw produce sells in the market;" and, therefore, in an inquiry into the causes which may elevate the price of raw produce, compared with the cost of production, we are inquiring into the causes which may elevate rent.

In reference to the first cause which Mr. Malthus has assigned for the rise of rent, namely, "that quality of the earth by which it can be made to yield a greater portion of the necessaries of life than is required for the maintenance of the persons employed on the land," he makes the following observations: "We still want to know why the consumption and supply are such as to make the price so greatly exceed the cost of production, and the main cause is evidently the *fertility* of the earth in producing the necessaries of life. Diminish this plenty, diminish the fertility of the soil, and the excess will diminish; diminish it still further, and it will disappear." True, the excess of necessaries will diminish and disappear, but that is not the question. The question is, whether the excess of their price above the cost of their production will diminish and disappear, for it is on this that money rent depends. Is Mr. Malthus warranted in his inference, that because the excess of quantity will diminish and disappear, therefore " the cause of the *high price* of the necessaries of life above the cost of production is to be found in their abundance, rather than in their scarcity, and is not only essentially different from the high price occasioned by artificial monopolies, but from the high price of those peculiar products of the earth, not connected with food, which may be called natural and necessary monopolies? "

Are there no circumstances under which the fertility of the land and the plenty of its produce may be diminished without occasioning a diminished excess of its price above the cost of production, that is to say, a diminished rent? If there are, Mr. Malthus's proposition is much too universal; for he appears to me to state it as a general principle, true under all circumstances, that rent will rise with the increased fertility of the land, and will fall with its diminished fertility.

Mr. Malthus would undoubtedly be right if, of any given farm, in proportion as the land yielded abundantly, a greater share of the whole produce were paid to the landlord; but the contrary is the fact; when no other but the most fertile land is in cultivation, the landlord has the smallest proportion of the whole produce, as well as the smallest value, and it is only when inferior lands are required to feed an augmenting population that both the landlord's share of the whole produce and the value he receives progressively increase.

Suppose that the demand is for a million of quarters of corn, and that they are the produce of the land actually in cultivation. Now, suppose the fertility of all the land to be so diminished that the very same lands will yield only 900,000 quarters. The demand being for a million of quarters, the price of corn would rise, and recourse must necessarily be had to land of an inferior quality sooner than if the superior land had continued to produce a million of quarters. But it is this necessity of taking inferior land into cultivation which is the cause of the rise of rent, and will elevate it, although the quantity of corn received by the landlord be reduced in quantity. Rent, it must be remembered, is not in proportion to the absolute fertility of the land in cultivation, but in proportion to its relative fertility. Whatever cause may drive capital to inferior land must elevate rent on the superior land; the cause of rent being, as stated by Mr. Malthus in his third proposition, " the comparative scarcity of the most fertile land." The price of corn will naturally rise with the difficulty of producing the last portions of it, and the value of the whole quantity produced on a particular farm will be increased, although its quantity be diminished; but as the cost of production will not increase on the more fertile land, as wages and profits taken together will continue always of the same value,[1] it is evident that the excess of price above the cost of production, or, in other words, rent, must rise with the diminished fertility of the land, unless it is counteracted by a great reduction of capital, population, and demand. It does not appear, then, that Mr. Malthus's proposition is correct: rent does not immediately and necessarily rise or fall with the increased or diminished fertility of the land; but its increased fertility renders it capable of paying at some future time an augmented rent. Land possessed of very little fertility can never bear any rent; land of moderate fertility may be made, as population increases, to bear a moderate rent; and land of great fertility a high rent; but it is one thing to be able to bear a high rent, and another thing actually to pay it. Rent may be lower in a country where lands are exceedingly fertile than in a country where they yield a moderate return, it being in proportion rather to relative than absolute fertility—to the value of the produce, and not to its abundance.[2]

Mr. Malthus supposes that the rent on land yielding those peculiar products of the earth which may be called natural and

[1] See page 70, where I have endeavoured to show that whatever facility or difficulty there may be in the production of corn, wages and profits together will be of the same value. When wages rise, it is always at the expense of profits, and when they fall, profits always rise.

[2] Mr. Malthus has observed in a late publication that I have misunderstood him in this passage, as he did not mean to say that rent immediately and necessarily rises and falls with the increased or diminished fertility of the land. If so, I certainly did misunderstand him. Mr. Malthus's words are, " Diminish this plenty, diminish the fertility of the soil, and the excess (rent) will diminish; diminish it still further, and it will disappear." Mr. Malthus does not state his proposition conditionally, but absolutely. I contended against what I understood him to maintain, that a diminution of the fertility of the soil was incompatible with an increase of rent.

necessary monopolies is regulated by a principle essentially different from that which regulates the rent of land that yields the necessaries of life. He thinks that it is the scarcity of the products of the first which is the cause of a high rent, but that it is the abundance of the latter which produces the same effect.

This distinction does not appear to me to be well founded; for you would as surely raise the rent of land yielding scarce wines, as the rent of corn land, by increasing the abundance of its produce, if, at the same time, the demand for this peculiar commodity increased; and without a similar increase of demand, an abundant supply of corn would lower instead of raise the rent of corn land. Whatever the nature of the land may be, high rent must depend on the high price of the produce; but, given the high price, rent must be high in proportion to abundance and not to scarcity.

We are under no necessity of producing permanently any greater quantity of a commodity than that which is demanded. If by accident any greater quantity were produced it would fall below its natural price, and therefore would not pay the cost of production, including in that cost the usual and ordinary profits of stock: thus the supply would be checked till it conformed to the demand, and the market price rose to the natural price.

Mr. Malthus appears to me to be too much inclined to think that population is only increased by the previous provision of food—" that it is food that creates its own demand "—that it is by first providing food that encouragement is given to marriage, instead of considering that the general progress of population is affected by the increase of capital, the consequent demand for labour, and the rise of wages; and that the production of food is but the effect of that demand.

It is by giving the workmen more money, or any other commodity in which wages are paid, and which has not fallen in value, that his situation is improved. The increase of population and the increase of food will generally be the effect, but not the necessary effect, of high wages. The amended condition of the labourer, in consequence of the increased value which is paid him, does not necessarily oblige him to marry and take upon himself the charge of a family—he will, in all probability, employ a portion of his increased wages in furnishing himself abundantly with food and necessaries—but with the remainder he may, if it please him, purchase any commodities that may contribute to his enjoyments—chairs, tables, and hardware; or better clothes, sugar, and tobacco. His increased wages, then, will be attended with no other effect than an increased demand for some of those commodities; and as the race of labourers will not be materially increased, his wages will continue permanently high. But although this might be the consequence of high wages, yet so great are the delights of domestic society, that, in practice, it is invariably found that an increase of population follows the amended condition of the labourer; and it is only because it does so, that, with the trifling exception already mentioned, a new and increased demand arises for food. This demand, then, is the effect of an increase of capital and population, but not the cause—it is only because the expenditure of the people takes this direction, that the market price of necessaries exceeds the natural price, and that the quantity of food required is produced; and it is because the number of people is increased that wages again fall.

What motive can a farmer have to produce more corn than is actually demanded, when the consequence would be a depression of its market price below its natural price, and consequently a privation to him of a portion of his profits, by reducing them below the general rate? " If," says Mr. Malthus, " the necessaries of life, the most important products of land, had not the property of creating an increase of demand proportioned to their increased quantity, such increased quantity would occasion a fall in their exchangeable value.[1] However abundant might be the produce of the country, its population might remain stationary; and this abundance without a proportionate demand, and with a very high corn price of labour, which would naturally take place under these circumstances, might reduce the price of raw produce, like the price of manufactures, to the cost of production."

Might reduce the price of raw produce to the cost of production. Is it ever for any length of time either above or below this price? Does not Mr. Malthus himself state it never to be so? " I hope," he says, " to be excused for dwelling a little, and presenting to the reader, in various forms, the doctrine that corn, in reference to the quantity *actually produced*, is sold at its necessary price like manufactures, because I consider it as a truth of the highest importance, which has been overlooked by the economists, by Adam Smith, and all those writers, who have represented raw produce as selling always at a monopoly price."

[1] Of what increased quantity does Mr. Malthus speak? Who is to produce it? Who can have any motive to produce it before any demand exists for an additional quantity?

" Every extensive country may thus be considered as possessing a gradation of machines for the production of corn and raw materials, including in this gradation not only all the various qualities of poor land, of which every territory has generally an abundance, but the inferior machinery, which may be said to be employed when good land is further and further forced for additional produce. As the price of raw produce continues to rise, these inferior machines are successively called into action; and as the price of raw produce continues to fall, they are successively thrown out of action. The illustration here used serves to show at once the *necessity of the actual price of corn to the actual produce*, and the different effect which would attend a great reduction in the price of any particular manufacture, and a great reduction in the price of raw produce." [1]

How are these passages to be reconciled to that which affirms, that if the necessaries of life had not the property of creating an increase of demand proportioned to their increased quantity, the abundant quantity produced would then, and then only, reduce the price of raw produce to the cost of production? If corn is never under its natural price, it is never more abundant than the actual population require it to be for their own consumption; no store can be laid up for the consumption of others; it can never, then, by its cheapness and abundance, be a stimulus to population. In proportion as corn can be produced cheaply, the increased wages of the labourers will have more power to maintain families. In America population increases rapidly because food can be produced at a cheap price, and not because an abundant supply has been previously provided. In Europe population increases comparatively slowly, because food cannot be produced at a cheap value. In the usual and ordinary course

[1] *Inquiry*, etc. " In all progressive countries the average price of corn is never higher than what is necessary to continue the average increase of produce."—*Observations*, p. 21.

" In the employment of fresh capital upon the land, to provide for the wants of an increasing population, whether this fresh capital is employed in bringing more land under the plough, or improving land already in cultivation, the main question always depends upon the expected returns of this capital; and no part of the gross profits can be diminished without diminishing the motive to this mode of employing it. Every diminution of price not fully and immediately balanced by a proportionate fall in all the necessary expenses of a farm, every tax on the land, every tax on farming stock, every tax on the necessaries of farmers, will tell in the computation; and if, after all these outgoings are allowed for, the price of the produce will not leave a fair remuneration for the capital employed, according to the general rate of profits, and a rent at least equal to the rent of the land in its former state, no sufficient motive can exist to undertake the projected improvement."—*Observations*, p. 22.

of things the demand for all commodities precedes their supply.
By saying that corn would, like manufactures, sink to its price
of production, if it could not raise up demanders, Mr. Malthus
cannot mean that all rent would be absorbed; for he has
himself justly remarked that if all rent were given up by the
landlords corn would not fall in price; rent being the effect
and not the cause of high price, and there being always one
quality of land in cultivation which pays no rent whatever,
the corn from which replaces by its price only wages and
profits.

In the following passage, Mr. Malthus has given an able
exposition of the causes of the rise in the price of raw produce
in rich and progressive countries, in every word of which I concur;
but it appears to me to be at variance with some of the propositions maintained by him in his essay on rent. " I have no
hesitation in stating that, independently of the irregularities
in the currency of a country, and other temporary and accidental
circumstances, the cause of the high comparative money price
of corn is its high comparative *real price*, or the greater quantity
of capital and labour which must be employed to produce it;
and that the reasons why the real price of corn is higher, and
continually rising in countries which are already rich and still
advancing in prosperity and population, is to be found in the
necessity of resorting constantly to poorer land, to machines
which require a greater expenditure to work them, and which
consequently occasion each fresh addition to the raw produce
of the country to be purchased at a greater cost; in short, it
is to be found in the important truth that corn in a progressive
country is sold at a price necessary to yield the actual supply;
and that, as this supply becomes more and more difficult, the
price rises in proportion."

The real price of a commodity is here properly stated to depend
on the greater or less quantity of labour and capital (that is,
accumulated labour) which must be employed to produce it.
Real price does not, as some have contended, depend on money
value; nor, as others have said, on value relatively to corn,
labour, or any other commodity taken singly, or to all commodities collectively; but, as Mr. Malthus justly says, " on the
greater (or less) quantity of capital and labour which must be
employed to produce it."

Among the causes of the rise of rent, Mr. Malthus mentions,
" such an increase of population as will lower the wages of labour."
But if, as the wages of labour fall, the profits of stock rise, and they

be together always of the same value,[1] no fall of wages can raise rent, for it will neither diminish the portion nor the value of the portion of the produce which will be allotted to the farmer and labourer together; and, therefore, will not leave a larger portion nor a larger value for the landlord. In proportion as less is appropriated for wages, more will be appropriated for profits, and *vice versa*. This division will be settled by the farmer and his labourers without any interference of the landlord; and, indeed, it is a matter in which he can have no interest, otherwise than as one division may be more favourable than another, to new accumulations, and to a further demand for land. If wages fell, profits, and not rent, would rise. If wages rose, profits, and not rent, would fall. The rise of rent and wages, and the fall of profits, are generally the inevitable effects of the same cause— the increasing demand for food, the increased quantity of labour required to produce it, and its consequently high price. If the landlord were to forego this whole rent, the labourers would not be in the least benefited. If it were possible for the labourers to give up their whole wages, the landlords would derive no advantage from such a circumstance; but in both cases the farmers would receive and retain all which they relinquish. It has been my endeavour to show in this work that a fall of wages would have no other effect than to raise profits. Every rise of profits is favourable to the accumulation of capital, and to the further increase of population, and therefore would, in all probability, ultimately lead to an increase of rent.

Another cause of the rise of rent, according to Mr. Malthus, is " such agricultural improvements or such increase of exertions as will diminish the number of labourers necessary to produce a given effect." To this passage I have the same objection that I had against that which speaks of the increased fertility of land being the cause of an immediate rise of rent. Both the improvement in agriculture, and the superior fertility, will give to the land a capability of bearing at some future period a higher rent, because with the same price of food there will be a great additional quantity; but till the increase of population be in the same proportion, the additional quantity of food would not be required, and, therefore, rents would be lowered and not raised. The quantity that could under the then existing circumstances be consumed could be furnished either with fewer hands, or with a less quantity of land, the price of raw produce would fall, and capital would be withdrawn from the land.[2] Nothing can

[1] See p. 72.   [2] See page 44, etc.

raise rent but a demand for new land of an inferior quality, or some cause which shall occasion an alteration in the relative fertility of the land already under cultivation.[1] Improvements in agriculture, and in the division of labour, are common to all land; they increase the absolute quantity of raw produce obtained from each, but probably do not much disturb the relative proportions which before existed between them.

Mr. Malthus has justly commented on the error of Dr. Smith's argument, that corn is of so peculiar a nature that its production cannot be encouraged by the same means that the production of all other commodities is encouraged. He observes, " It is by no means intended to deny the powerful influence of the price of corn upon the price of labour, on an average of a considerable number of years; but that this influence is not such as to prevent the movement of capital to or from the land, which is the precise point in question, will be made sufficiently evident by a short inquiry into the manner in which labour is paid and brought into the market, and by a consideration of the consequences to which the assumption of Adam Smith's proposition would inevitably lead." [2]

Mr. Malthus then proceeds to show that demand and high price will as effectually encourage the production of raw produce as the demand and high price of any other commodity will encourage its production. In this view it will be seen, from what I have said of the effects of bounties, that I entirely concur. I have noticed the passage from Mr. Malthus's *Observations on the Corn Laws*, for the purpose of showing in what a different

[1] It is not necessary to state on every occasion, but it must be always understood, that the same results will follow, as far as regards the price of raw produce and the rise of rents, whether an additional capital of a given amount be employed on new land, for which no rent is paid, or on land already in cultivation, if the produce obtained from both be precisely the same in quantity.—See p. 37.

M. Say, in his notes to the French translation of this work, has endeavoured to show that there is not at any time land in cultivation which does not pay a rent, and having satisfied himself on this point, he concludes that he has overturned all the conclusions which result from that doctrine. He infers, for example, that I am not correct in saying that taxes on corn and other raw produce, by elevating their price, fall on the consumer, and do not fall on rent. He contends that such taxes must fall on rent. But before M. Say can establish the correctness of this inference, he must also show that there is not any capital employed on the land for which no rent is paid (see the beginning of this note, and pages 33 and 38 of the present work); now this he has not attempted to do. In no part of his notes has he refuted or even noticed that important doctrine. By his note to page 182 of the second volume of the French edition, he does not appear to be aware that it has even been advanced.

[2] *Observations on the Corn Laws*, p. 4.

sense the term real price is used here, and in his other pamphlet, entitled *Grounds of an Opinion,* etc. In this passage Mr. Malthus tells us that " it is clearly an increase of real price alone which can encourage the production of corn," and, by real price, he evidently means the increase in its value relatively to all other things, or, in other words, the rise in its market above its natural price, or the cost of its production. If by real price this is what is meant, although I do not admit the propriety of thus naming it, Mr. Malthus's opinion is undoubtedly correct; it is the rise in the market price of corn which alone encourages its production; for it may be laid down as a principle uniformly true that the only great encouragement to the increased production of a commodity is its market value exceeding its natural or necessary value.

But this is not the meaning which Mr. Malthus, on other occasions, attaches to the term real price. In the essay on rent Mr. Malthus says, by " the real growing price of corn I mean the real *quantity* of labour and capital *which has been employed* to produce the last additions which have been made to the national produce." In another part he states " the cause of the high comparative real price of corn to be the greater *quantity* of capital and labour which must be *employed* to produce it." [1] Suppose that, in the foregoing passage, we were to substitute this definition of real price, would it not then run thus?—" It is clearly the increase in the quantity of labour and capital which must be employed to produce corn, which alone can encourage its production." This would be to say, that it is clearly the rise in the natural or necessary price of corn which encourages its production—a proposition which could not be maintained. It is not the price at which corn can be produced that has any influence on the quantity produced, but the price at which it can be sold. It is in proportion to the degree of the difference of its price above or below the cost of production that capital is attracted to or repelled from the land. If that excess be such as to give to capital so employed a greater than the general profit of stock, capital will go to the land; if less, it will be withdrawn from it.

It is not, then, by an alteration in the real price of corn that its production is encouraged, but by an alteration in its market

---

[1] Upon showing this passage to Mr. Malthus, at the time when these papers were going to the press, he observed, " that in these two instances he had inadvertently used the term *real price,* instead of *cost of production.* It will be seen, from what I have already said, that to me it appears that in these two instances he has used the term *real price* in its true and just acceptation, and that in the former case only it is incorrectly applied.

price. It is not "because a greater quantity of capital and labour must be employed to produce it (Mr. Malthus's just definition of real price) that more capital and labour are attracted to the land, but because the market price rises above this, its real price, and, notwithstanding the increased charge, makes the cultivation of land the more profitable employment of capital."

Nothing can be more just than the following observations of Mr. Malthus on Adam Smith's standard of value. "Adam Smith was evidently led into this train of argument from his habit of considering *labour as the standard measure of value* and corn as the measure of labour. But that corn is a very inaccurate measure of labour the history of our own country will amply demonstrate; where labour, compared with corn, will be found to have experienced very great and striking variations, not only from year to year, but from century to century, and for ten, twenty, and thirty years together. *And that neither labour nor any other commodity can be an accurate measure of real value in exchange* is now considered as one of the most incontrovertible doctrines of political economy, and, indeed, follows from the very definition of value in exchange."

If neither corn nor labour are accurate measures of real value in exchange, which they clearly are not, what other commodity is?—certainly none. If, then, the expression, real price of commodities, have any meaning, it must be that which Mr. Malthus has stated in the essay on rent—it must be measured by the proportionate quantity of capital and labour necessary to produce them.

In Mr. Malthus's *Inquiry into the Nature of Rent*, he says, "that, independently of irregularities in the currency of a country, and other temporary and accidental circumstances, the cause of the high comparative money price of corn is its high comparative real price, *or the greater quantity of capital and labour which must be employed to produce it.*" [1]

This, I apprehend, is the correct account of all permanent variations in price, whether of corn or of any other commodity. A commodity can only permanently rise in price either because a greater quantity of capital and labour must be employed to produce it, or because money has fallen in value; and, on the contrary, it can only fall in price, either because a less quantity of capital and labour may be employed to produce it, or because money has risen in value.

[1] Page 40.

A variation arising from the latter of these alternatives, an altered value of money, is common at once to all commodities; but a variation arising from the former cause is confined to the particular commodity requiring more or less labour in its production. By allowing the free importation of corn, or by improvements in agriculture, raw produce would fall; but the price of no other commodity would be affected, except in proportion to the fall in the real value, or cost of production, of the raw produce which entered into its composition.

Mr. Malthus, having acknowledged this principle, cannot, I think, consistently maintain that the whole money value of all the commodities in the country must sink exactly in proportion to the fall in the price of corn. If the corn consumed in the country were of the value of 10 millions per annum, and the manufactured and foreign commodities consumed were of the value of 20 millions, making altogether 30 millions, it would not be admissible to infer that the annual expenditure was reduced to 15 millions because corn had fallen 50 per cent., or from 10 to 5 millions.

The value of the raw produce which entered into the composition of these manufactures might not, for example, exceed 20 per cent. of their whole value, and, therefore, the fall in the value of manufactured commodities, instead of being from 20 to 10 millions, would be only from 20 to 18 millions; and after the fall in the price of corn of 50 per cent., the whole amount of the annual expenditure, instead of falling from 30 to 15 millions, would fall from 30 to 23 millions.[1]

This, I say, would be their value if you supposed it possible that with such a cheap price of corn no more corn and commodities would be consumed; but as all those who had employed capital in the production of corn on those lands which would no longer be cultivated could employ it in the production of manufactured goods, and only a part of those manufactured goods would be given in exchange for foreign corn, as on any other supposition no advantage would be gained by importation and low prices, we should have the additional value of all that quantity of manufactured goods which were so produced and not exported to add to the above value, so that the real diminution, even in money value, of all the commodities in the country,

---

[1] Manufactures, indeed, could not fall in any such proportion, because, under the circumstances supposed, there would be a new distribution of the precious metals among the different countries. Our cheap commodities would be exported in exchange for corn and gold, till the accumulation of gold should lower its value and raise the money price of commodities.

corn included, would be equal only to the loss of the landlords, by the reduction of their rents, while the quantity of objects of enjoyment would be greatly increased.

Instead of thus considering the effect of a fall in the value of raw produce, as Mr. Malthus was bound to do by his previous admission, he considers it as precisely the same thing as a rise of 100 per cent. in the value of money, and, therefore, argues as if all commodities would sink to half their former price.

" During the twenty years beginning with 1794," he says, " and ending with 1813, the average price of British corn per quarter was about 83 shillings; during the ten years ending with 1813, 92 shillings; and during the last five years of the twenty, 108 shillings. In the course of these twenty years, the government borrowed near 500 millions of real capital; for which, on a rough average, exclusive of the sinking fund, it engaged to pay about 5 per cent. But if corn should fall to 50 shillings a quarter, and other commodities in proportion, instead of an interest of about 5 per cent., the government would really pay an interest of 7, 8, 9, and, for the last 200 millions, 10 per cent.

" To this extraordinary generosity towards the stockholders I should be disposed to make no kind of objection, if it were not necessary to consider by whom it is to be paid; and a moment's reflection will show us that it can only be paid by the industrious classes of society and the landlords, that is, by all those whose nominal income will vary with the variations in the measure of value. The nominal revenues of this part of the society, compared with the average of the last five years, will be diminished one half, and out of this nominally reduced income they will have to pay the same nominal amount of taxes." [1]

In the first place, I think I have already shown that even the value of the gross income of the whole country will not be diminished in the proportion for which Mr. Malthus here contends; it would not follow that because corn fell 50 per cent. each man's gross income would be reduced 50 per cent. in value; [2] his net income might be actually increased in value.

In the second place, I think the reader will agree with me that the increased charge, if admitted, would not fall exclusively " on the landlords and the industrious classes of society; " the stockholder, by his expenditure, contributes his share to the support

[1] *The Grounds of an Opinion*, etc., p. 36.
[2] Mr. Malthus, in another part of the same work, supposes commodities to vary 25 or 20 per cent. when corn varies $33\frac{1}{3}$.

of the public burdens in the same way as the other classes of society. If, then, money became really more valuable, although he would receive a greater value, he would also pay a greater value in taxes, and, therefore, it cannot be true that the whole addition to the real value of the interest would be paid by " the landlords and the industrious classes."

The whole argument, however, of Mr. Malthus, is built on an infirm basis: it supposes, because the gross income of the country is diminished, that, therefore, the net income must also be diminished in the same proportion. It has been one of the objects of this work to show that, with every fall in the real value of necessaries, the wages of labour would fall, and that the profits of stock would rise; in other words, that of any given annual value a less portion would be paid to the labouring class, and a larger portion to those whose funds employed this class. Suppose the value of the commodities produced in a particular manufacture to be £1000, and to be divided between the master and his labourers in the proportion of £800 to labourers and £200 to the master; if the value of these commodities should fall to £900, and £100 be saved from the wages of labour, in consequence of the fall of necessaries, the net income of the master would be in no degree impaired, and, therefore, he could with just as much facility pay the same amount of taxes after as before the reduction of price.[1]

It is of importance to distinguish clearly between gross revenue and net revenue, for it is from the net revenue of a society that all taxes must be paid. Suppose that all the commodities in the country, all the corn, raw produce, manufactured goods, etc., which could be brought to market in the course of the year, were of the value of 20 millions, and that in order to obtain this value the labour of a certain number of men was necessary, and that the absolute necessaries of these labourers required an expenditure of 10 millions; I should say that the gross revenue of such society was 20 millions, and its net revenue 10 millions. It does not follow from this supposition that the labourers should receive only 10 millions for their labour; they might receive 12, 14, or 15 millions, and in that case they would have

[1] Of net produce and gross produce M. Say speaks as follows: " The whole value produced is the gross produce; this value, after deducting from it the cost of production, is the net produce."—Vol. ii. p. 491. There can, then, be no net produce, because the cost of production, according to M. Say, consists of rent, wages, and profits. In page 508 he says, " The value of a product, the value of a productive service, the value of the cost of production, are all, then, similar values, whenever things are left to their natural course." Take a whole from a whole and nothing remains.

2, 4, or 5 millions of the net income. The rest would be divided between landlords and capitalists; but the whole net income would not exceed 10 millions. Suppose such a society paid 2 millions in taxes, its net income would be reduced to 8 millions.

Suppose now money to become more valuable by one-tenth, all commodities would fall, and the price of labour would fall, because the absolute necessaries of the labourer formed a part of those commodities, consequently the gross income would be reduced to 18 millions and the net income to 9 millions. If the taxes fell in the same proportion, and, instead of 2 millions, £1,800,000 only were raised, the net income would be further reduced to £7,200,000, precisely of the same value as the 8 millions were before, and therefore the society would neither be losers nor gainers by such an event. But suppose that after the rise of money, 2 millions were raised for taxes as before, the society would be poorer by £200,000 per annum, their taxes would be really raised one-ninth. To alter the money value of commodities, by altering the value of money, and yet to raise the same money amount by taxes, is then undoubtedly to increase the burthens of society.

But suppose of the 10 millions net revenue the landlords received five millions as rent, and that by facility of production, or by the importation of corn, the necessary cost of that article in labour was reduced 1 million, rent would fall 1 million, and the prices of the mass of commodities would also fall to the same amount, but the net revenue would be just as great as before; the gross income would, it is true, be only 19 millions, and the necessary expenditure to obtain it 9 millions, but the net income would be 10 millions. Now, suppose 2 millions raised in taxes on this diminished gross income, would the society altogether be richer or poorer? Richer, certainly; for after the payment of their taxes, they would have, as before, a clear income of 8 millions to bestow on the purchase of commodities, which had increased in quantity, and fallen in price, in the proportion of 20 to 19; not only then could the same taxation be endured, but greater, and yet the mass of the people be better provided with conveniences and necessaries.

If the net income of the society, after paying the same money taxation, be as great as before, and the class of landholders lose 1 million from a fall of rent, the other productive classes must have increased money incomes, notwithstanding the fall of prices. The capitalist will then be doubly benefited; the corn and butcher's meat consumed by himself and his family will be reduced in price; and the wages of his menial servants, of his gardeners, and labourers of all descriptions, will be also lowered. His horses and cattle will cost less, and be supported at a less expense. All the commodities in which raw produce enters as a principal part of their value will fall. This aggregate amount of savings, made on the expenditure of income, at the same time that his money income is increased, will then be doubly beneficial to him, and will enable him not only to add to his enjoyments, but to bear additional taxes, if they should be required: his additional consumption of taxed commodities will much more than make up for the diminished demand of landlords, consequent on the reduction of their rents. The same observations apply to farmers and traders of every description.

But it may be said that the capitalist's income will not be increased; that the million deducted from the landlord's rent will be paid in additional wages to labourers! Be it so; this will make no difference in the argument: the situation of the society will be improved, and they will be able to bear the same money burthens with greater facility than before; it will only prove what is still more desirable, that the situation of another class, and by far the most important class in society, is the one which is chiefly benefited by the new distribution. All that they receive more than 9 millions forms part of the net income of the country, and it cannot be expended without adding to its revenue, its happiness, or its power. Distribute, then, the net income as you please. Give a little more to one class and a little less to another, yet you do not thereby diminish it; a greater amount of commodities will be still produced with the same labour, although the amount of the gross money value of such commodities will be diminished; but the net money income of the country, that fund from which taxes are paid and enjoyments procured, would be much more adequate than before to maintain the actual population, to afford it enjoyments and luxuries, and to support any given amount of taxation.

That the stockholder is benefited by a great fall in the value of corn cannot be doubted; but if no one else be injured, that is no reason why corn should be made dear; for the gains of the stockholder are national gains, and increase, as all other gains do, the real wealth and power of the country. If they are unjustly benefited, let the degree in which they are so be accurately ascertained, and then it is for the legislature to devise a remedy; but no policy can be more unwise than to shut ourselves out from the great advantages arising from cheap corn, and abundant productions, merely because the stockholder would have an undue proportion of the increase.

To regulate the dividends on stock by the money value of corn has never yet been attempted. If justice and good faith required such a regulation, a great debt is due to the old stockholders; for they have been receiving the same money dividends for more than a century, although corn has, perhaps, been doubled or trebled in price.

But it is a great mistake to suppose that the situation of the stockholder will be more improved than that of the farmer, the manufacturer, and the other capitalists of the country; it will, in fact, be less improved.

The stockholder will undoubtedly receive the same money dividend, while not only the price of raw produce and labour fell, but the prices of many other things into which raw produce entered as a component part. This, however, is an advantage, as I have just stated, which he would enjoy in common with all other persons who had the same money incomes to expend:—his money income would not be increased; that of the farmer, manufacturer, and other employers of labour would, and consequently they would be doubly benefited.

It may be said that, although it may be true that capitalists would be benefited by a rise of profits, in consequence of a fall of wages, yet that their incomes would be diminished by the fall in the money value of their commodities. What is to lower them? Not any alteration in the value of money for nothing has been supposed to occur to alter the value of money. Not any diminution in the quantity of labour necessary to produce their commodities, for no such cause has operated, and if it did operate, would not lower money profits, though it might lower money prices. But the raw produce of which commodities are made is supposed to have fallen in price, and, therefore, commodities will fall on that account. True, they will fall, but their fall will not be attended with any diminution in the money income of the producer. If he sell his commodity for less money, it is only because one of the materials from which it is made has fallen in value. If the clothier sell his cloth for £900 instead of £1000, his income will not be less, if the wool from which it is made has declined £100 in value.

Mr. Malthus says, "It is true that the last additions to the agricultural produce of an improving country are not attended with a large proportion of rent; and it is precisely this circumstance that may make it answer to a rich country to import some of its corn, if it can be secure of obtaining an equable supply. But in all cases the importation of foreign corn must fail to answer nationally if it is not so much cheaper than the corn that can be grown at home as to equal both the profits and the rent of the grain which it displaces."—*Grounds*, etc., p. 36.

In this observation Mr. Malthus is quite correct; but imported corn *must* be always so much cheaper than the corn that can be grown at home, " as to equal both the profits and the rent of the grain which it displaces." If it were not, no advantage to any one could be obtained by importing it.

As rent is the effect of the high price of corn, the loss of rent is the effect of a low price. Foreign corn never enters into competition with such home corn as affords a rent; the fall of price invariably affects the landlord till the whole of his rent is absorbed;—if it fall still more, the price will not afford even the common profits of stock; capital will then quit the land for some other employment, and the corn which was before grown upon it will then, and not till then, be imported. From the loss of rent there will be a loss of value, of estimated money value, but there will be a gain of wealth. The amount of the raw produce and other productions together will be increased; from the greater facility with which they are produced they will, though augmented in quality, be diminished in value.

Two men employ equal capitals—one in agriculture, the other in manufactures. That in agriculture produces a net annual value of £1200, of which £1000 is retained for profit and £200 is paid for rent; the other in manufactures produces only an annual value of £1000. Suppose that, by importation, the same quantity of corn which cost £1200 can be obtained for commodities which cost £950, and that, in consequence, the capital employed in agriculture is diverted to manufactures, where it can produce a value of £1000, the net revenue of the country will be of less value, it will be reduced from £2200 to £2000; but there will not only be the same quantity of commodities and corn for its own consumption, but also as much addition to that quantity as £50 would purchase, the difference between the value at which its manufactures were sold to the foreign country and the value of the corn which was purchased from it.

Now this is precisely the question respecting the advantage of importing or growing corn; it never can be imported till the quantity obtained from abroad by the employment of a given capital exceeds the quantity which the same capital will enable us to grow at home—exceeds not only that quantity which falls to the share of the farmer, but also that which is paid as rent to the landlord.

Mr. Malthus says, " It has been justly observed by Adam Smith that no equal quantity of productive labour employed in manufactures can ever occasion so great a reproduction as in agriculture." If Adam Smith speaks of value, he is correct; but if he speaks of riches, which is the important point, he is mistaken; for he has himself defined riches to consist of the necessaries, conveniences, and enjoyments of human life. One set of necessaries and conveniences admits of no comparison with another set; value in use cannot be measured by any known standard; it is differently estimated by different persons.

INDEX

AGRICULTURE, effect of taxation on, 105, 118, 220
    effect of war on, 177, 181
    improved methods in, 40, 42, 71
    nature's aids to, 39
    in relation to productive labour in general, 234-237

Banks, 238
    rate of interest on money lent by, 246, 247
    system of exchange with the Mint, 241, 242
Barton, on capital and labour, 270
Bills of exchange, 90, 91, 92
    in relation to foreign currencies, 85, 92
Bounties, effect on colonial trade, 227
    effect on price of commodities, 202
    effect on relative value of money, 87, 207, 212
    on exportation of corn, 201
    on manufactures, 209
    on production, 215
Buchanan, on bounties on exportation of corn, 211
    on depreciation of the currency, 240, 251
    on derivation of rent, 40, 224
    on monopoly prices of raw produce, 166
    on Poor Laws, 61
    on rent as form of transferred revenue, 272
    on rent in proportion to production, 224
    on tax on malt, 168
    on taxes on wages, 140, 143, 260
Bullion, exports and imports of, 150

Capital, 13
    accumulation of, 73, 74, 79
    British, in colonial trade, 231
    circulating, 19, 23, 49, 94, 134
    durability of, 18, 24, 27, 94
    emigration of, 83
    employment of, 48-51, 70, 164, 234, 235
    fixed, 23, 29, 94, 133
    increase in, causes of, 53, 54, 155
    in foreign trade, 77, 78, 236
    in home trade, 81, 83, 135
    invested in land, 36, 41, 178-180
    invested in machinery, 25, 26, 265
    national, 94
    nature of, 53
    portion yielding no rent, 36, 37, 38, 64, 166, 189
    price of commodities in relation to, 284
    rent of land in relation to, 36
    taxes on, 95
Cash accounts, 248
Colonial currency, 240

Colonial trade, 227
       effect on British profits, 231
       monopoly of, 231, 232
       preference in, 231
Commercial treaties, 228
Corn, bounties on, 201
       comparative value in poor and rich countries, 253, 254
       effect of increased prices of, 66, 68, 72, 74, 102, 202, 225, 226
       effect of war on, 178-181
       fluctuations in value of, 7, 8
       importation of, 291
       laws regulating price of, 38, 64
       price of labour regulated by, 211, 282, 284
       prohibitions on importation of, 209, 210
       rent in relation to price of, 209, 291
       rent produced by production of, 38, 39, 40, 44
       standard of price of all commodities, 283, 284, 285, 286
       taxation on, 99, 104, 105, 113, 122, 123, 132
       value as capital compared with cotton, 20, 21
       wages in relation to, 64
Corn rent, 58, 99
Cotton trade, 20, 21
Currency, colonial, 240
       depreciation of, 92, 135, 207, 239, 240
       in reign of William III., 250
       laws regulating value of, 238
       system proposed for security of, 241
       variations in value of, 248

Decker, Sir Matthew, on taxes, 153
Demand and supply, influence on prices, 260
       principles of, 49

Edinburgh Review, article on bounties on corn, 202, 206
Exchequer Bills, 198, 199
Exportation, bounties on, 201
       effect of accumulation of money at home on, 212

Farmer, distribution of capital of, 36
       profits of, 65-68
Fixed capital, 23, 29, 94, 133
Foreign trade, 77
       effect of bounties on, 218
       employment of capital in, 77, 78, 236
       general and reciprocal advantages of, 81, 82
       home profits in relation to, 78, 80
       money transactions in, 83
       value of money affected by, 230
Free trade, 218, 231
       in relation to the colonies, 227

Gold, assumed uniform value of, 47
       comparative value in different countries of, 89, 253, 256
       distribution in commerce, 83
       effect of forced abundance of, 107, 150
       fluctuating purchasing value of, 7, 10, 47, 238, 248
       standard measure of value, 28, 29
       taxes on, 122, 128, 132, 133
Government stock, interest on, 198, 199

Gross produce, 265, 266, 267
Gross revenue, 234, 268, 287
Ground rents, 129, 131

Home trade, 81, 236
Houses, taxes on, 129

Importation, prohibition of, 201, 209, 210
Interest, diminished by accumulation of capital, 192.   *See also* Profits
    variations in rate of, 198
Income, national, 237
    taxes on, 95, 135

Labour, capital in relation to, 13
    comparative value in poor and rich countries, 253
    constancy of supply of, 104, 105
    decreased profits caused by rise in price of, 21
    determining factor in value of all commodities, 6, 7, 14, 15, 16, 17,
        21, 22, 28-30, 31, 37, 48, 284
    different qualities of, 11
    economy in, 15
    effect of taxes on wages on, 142, 144
    machinery in relation to, 18, 26, 31, 264, 265-267
    market price of, 53, 54
    natural price of, 54
    nature's aids to, 190
    price of corn in relation to, 211, 282, 284
    profits in relation to quantity of, 76
    rent of land in relation to, 37, 46
Land, capital in relation to, 36, 179, 180
    degrees of fertility, 35, 42
    investment of capital in, 234-237
    laws regulating rent of, 219-226
Landlord, interests of, 225
    profits of, 72, 75. 136, 137, 221
    share of raw produce as rent, 35, 36, 43
    taxes paid by, 116, 146, 147, 153, 166
Land-tax, 115, 137
    effect on cultivation and production, 118, 120
Lauderdale, Lord, on alterations in value of commodities, 261
    on increase of riches by monopoly, 184
    on regulation of the currency, 251
Legacy duty, 96
Locke, on improvement of currency, 250
Luxuries, taxes on, 153, 158, 159

Machinery, effect on gross and net produce, 266, 267
        on population, 266
        on price of commodities, 26, 271
        on production, 22, 23, 31, 42, 139, 182
        on value of labour, 8, 9, 18
    improvements in, relating to agriculture, 42
    influence on interests of different classes, 263
Malthus, on cost as opposed to value of commodities, 30
    on fluctuations in values, 10, 11
    on Poor Laws, 61
    on price of labour in relation to population, 142
    on rent, 272-292

Manufacturer, fixed capital of, 25, 26
    profits of, 65, 69
Manufactures, bounties on, 209
    nature's aids to, 40
    sudden changes in demand and supply, 175
    taxes on, 258
Market price, 48
Melon, on National Debt, 161
Mercantile system, 212
Mines, effect of gold taxes on, 123-125
    laws regulating rent of, 46, 220
Mint, the, exchange of money with banks, 241
Money. *See also* Currency
    assumed invariability in value of, 64
    causes affecting relative value of, 87
    circulation of, 91
    comparative value in different countries, 87, 88, 89
    distribution in commerce, 83
    effect of forced abundance of, 150, 151, 152
    effect of gold tax on, 125, 126
    fluctuations in value of, 30, 47, 88, 90, 149, 151, 248, 284, 288
    transactions in foreign trade, 83, 87
    value in relation to price of commodities, 90
Monopoly, in colonial trade, 231, 232
    increase of riches by, 184
    injurious effects of, 229
    prices, 165

National debt, 161
    interest on, 199
National loans, 161-164
Natural price, 48
Navy Bills, 198
Net produce, influence of machinery on, 266, 267
Net revenue, 234, 268, 287

Overpopulation, remedies for, 56

Paper currency, 123, 238
    abuse of, 151
    regulation of, 241, 244
Pitt, on Poor Laws, 62
Poland, money values in, 256
Poor Laws, the, 61
    affecting the farmer, 171
    affecting the manufacturer, 173
    paid by consumer, 172
Population, corn supply in relation to, 103, 279
    improved agriculture in relation to, 42, 56
    machinery, effect of, 266
    regulated by capital, 41
Portugal, wine trade with England, 82-87
    gold mines of, 256
Preference in colonial trade, 231
Prices, influence of demand and supply on, 260
    natural and market, 48
    of necessaries, 102
Probate duty, 96

Production, bounties on, 215
    effect of increased, 194
    effect of land-taxes on, 118, 120
    fluctuations in, 48
    in relation to national capital, 94
    in relation to population, 56
Profits, affected by supply and demand, 50, 71
    causes of diminution in, 73, 74
    consumers' and producers', 81
    dependent on quantity of labour employed, 76
    effect of accumulation on, 192
    effect of bounties on production, 216
    effect of colonial trade on, 231
    fluctuations in, 70, 197
    foreign trade in relation to, 80
    laws regulating, 75
    permanent alterations in, 75
    on raw produce, 64, 66, 68, 72, 74
    taxes on, 132
    tendency to fall, 71, 73, 78
    unaffected by improvements or inventions, 81
    wages in relation to, 31, 64, 67, 70, 76, 80, 83
Prohibitions on importation of corn, 209, 210
    effect on prices of commodities, 212
Public Loans, 161-164, 198, 199

Rate of interest. _See_ Interest
Raw produce, in relation to price of commodities, 67, 69
    profits on, 64, 66, 68, 72, 74
    regulation of prices of, 165
    taxes on, 98, 104, 108
        paid by the consumer, 98, 100, 101, 104, 111, 118, 120, 122, 126
Rent, causes of rise of, 280
    capital in relation to, 36
    determined by relative fertility of land, 275
    effect of increasing population on, 40
    labour in relation to, 37
    landlord's share of increase, 58
    laws regulating, 34, 219, 220-226, 272
    lowered by improved agricultural methods. 42, 45
    Malthus on, 272-292
    of mines, laws regulating, 46
    nature of, 33, 36, 39
    price of corn in relation to, 209, 291
    Smith, Adam, on, 218-226
    taxes on, 110
    variations relative to production, 31
Revenue, distribution and beneficial employment of, 268, 289
    gross and net, 234, 268, 287
    taxes on, 95
Riches, dependent on facility of production, 185
    distinguished from value, 182
    increased by monopoly, 184
    increased by saving, 186
    objections to M. Say's doctrine of, 186-189
    standard of, 183

Salt tax, the, 155
Saving, from expenditure, 96, 106

Say, M., on commerce, 176
    on cost of production determining prices, 231, 261
    on effect of increased price of corn, 148
    on employment of capital, 164, 192, 193
                    on land, 234, 235
    on English land-tax, 119-121
    on foreign trade, 213
    on free productive powers of the earth, 35, 181
    on gross and net produce, 287
    on increased value of income, 186
    on interest on public loans, 199
    on over taxation, 155
    on payment of taxes by consumer and producer, 160, 258
    on produce by labour, 38
    on prohibition of importation, 212
    on rent, 189
    on Smith's theory of labour as standard measure of value, 190
    on taxation, 154, 156, 159, 169
    on taxes on transference of property, 97
    on value of silver, 183
Seignorage on coinage, 238, 252
Silver, alteration in value of, 183, 226, 239, 248
    currency in William III.'s reign, 250, 251
    distribution in commerce, 83
Simonde, on taxes paid by the producer, 258
Sismondi, on rent, 273
Smith, Adam, on accumulation of capital, 192
    on advantages derived from gross revenue, 234
    on bounties on exportation, 203, 204, 208
    on building and ground rent of houses, 129, 130, 131
    on colonial trade, 227, 231
    on commercial treaties, 228
    on comparative value of gold and silver in rich and poor countries, 253-256
    on corn as regulating money price of labour, 206
    on depreciation of silver in Spain, 149, 150, 207
    on desire for conveniences and ornaments, 195, 197
    on effect of low value of money, 226
    on employment of capital in carrying trade, 195
                    in home and foreign trade, 236
    on exportation of surplus produce, 193
    on fluctuations in purchasing power of labour, 9, 10, 11
    on fluctuations in profits, 197
    on labour as determining the price of commodities, 6, 7, 29, 190, 233
    on laws regulating rent of mines, 220, 221
    on nature's aids to agriculture, 39
    on paper currency, 239
    on profits due to foreign trade, 77, 78
    on profits in Holland, 193
    on rate of interest, 198
    on real and nominal price of labour, 183
    on relative value of gold and silver, 207
    on relative value of labour, 12, 13
    on rent, 33, 34, 41, 219, 223
    on riches and poverty, 182
    on Scotch system of cash accounts, 248
    standard of value criticised by Malthus, 284
    on taxation, 96, 115, 116, 117, 118
    on taxes on land, 116, 117

Smith, Adam—*continued*
  on taxes on malt, 167
  on taxes on necessaries, 153, 154
  on taxes on raw produce, 127
  on taxes on wages, 140, 145, 148, 149
  on value of agriculture, 40, 292
  on value in use, 5, 7, 187, 190
  on value of gold and silver, 248
  on variation in wages and profit, 13
Spanish gold mines, 256
  effect of gold tax on, 125-128
Stock, relative value of, 49
Stockholder, the position of, 289, 290
Supply and demand, fluctuations in, 50, 71

Taxes, 94
  abridgement of enjoyment caused by, 147, 157
  diminution of commodities caused by, 157
  effect on prices of commodities, 160
    on profits, 132
    on relative value of money, 87
  on capital, 95, 96, 97
  on commodities other than raw produce, 160
  on corn, 99, 104, 105, 132
  on expenditure or profits, 106
  on houses, 129
  on income, 95, 96
  on luxuries, 153, 158
  on money, 132, 133
  on necessaries, 153
  on profits, 132, 148
    in relation to the farmer, 135-137
    in relation to the manufacturer, 137
  on raw produce, 98, 104, 108, 109
    objections to, 100, 101, 105, 106, 107
    paid by consumer, 98, 100, 101, 104, 111, 118, 120, 122, 126
  on rent, 110
  on transference of property, 131
  on wages, 140, 142, 144, 148
  for war, 160, 161
  paid by consumer, 154, 155, 166, 167, 171
  paid by farmer, 166, 171
  paid by landlords, 146, 147, 153
  paid by producer, 259
  paid from net revenue, 287
Tithes, 112
Tobacco, 196
Torrens, Major, on corn trade, 181
  on variations in natural price of labour, 54
Tracy, M. Destutt de, on standard of values, 189
Trade, changes in channels of, 175
  foreign. *See* Foreign trade.
Treaties, commercial, 228
Turgot, reduction of market dues by, 156

Value, distinguished from cost, 30, 31
    from riches, 182
  fluctuations in, 9, 27

Value—*continued*
　　　in use, 187, 190
　　　of commodities, dependent on labour, 5, 6, 7, 14, 15, 16
　　　standards of, 184

Wages, affected by price of necessaries, 80
　　　effect of bounties on, 215
　　　effect of taxation on, 100, 105
　　　fall of profits caused by rise of, 21
　　　laws regulating, 57, 60
　　　natural price of, 52
　　　in relation to fixed capital, 29
　　　　　　　　to price of commodities, 57-61, 80
　　　　　　　　to profits, 31, 64, 67, 70, 76, 80, 81, 88
　　　　　　　　to value of money, 30
　　　relative value of, 31, 32
　　　taxes on, 140, 148
　　　variation in, 13, 27, 31, 55
Wealth, accumulation of, 79
　　　cause of increase of, 40
War, effect on trade, 176
　　　taxes for, 160, 161
Wine trade, profits in, 77, 79, 82, 84
Working classes, improvement in condition of, 57